Tributes
Volume 39

Word Recognition, Morphology and Lexical Reading
Essays in Honour of Cristina Burani

Volume 29
Computational Models of Rationality. Essays Dedicated to Gabriele Kern-Isberner on the Occasion of her 60th Birthday
Christoph Beierle, Gerhard Brewka and Matthias Thimm, eds.

Volume 30
Liber Amicorum Alberti. A Tribute to Albert Visser
Jan van Eijck, Rosalie Iemhoff and Joost J. Joosten, eds.

Volume 31
"Shut up," he explained. Essays in Honour of Peter K. Schotch
Gillman Payette, ed.

Volume 32
From Semantics to Dialectometry. Festschrift in Honour of John Nerbonne.
Martijn Wieling, Martin Kroon, Gertjan van Noord, and Gosse Bouma eds.

Volume 33
Logic and Computation. Essays in Honour of Amílcar Sernadas
Carlos Caleiro, Fransciso Dionísio, Paula Gouveia, Paulo Mateus and João Rasga, eds.

Volume 34
Models: Concepts, Theory, Logic, Reasoning, and Semantics. Essays Dedicated to Klaus-Dieter Schewe on the Occasion of his 60th Birthday
Atif Mashkoor, Qing Wang and Bernhrd Thalheim, eds.

Volume 35
Language, Evolution and Mind. Essays in Honour of Anne Reboul
Pierre Saint-Germier, ed.

Volume 36
Logic, Philosophy of Mathematics and their History.
Essays in Honor of W. W. Tait
Erich H. Reck, ed.

Volume 37
Argumentation-based Proofs of Endearment. Essays in Honor of Guillermo R. Simari on the Occasion of his 70th Birthday
Carlos I. Chesñevar, Marcelo A. Falappa, Eduardo Fermé, Alejandro J. García, Ana G. Maguitman, Diego C. Martínez, Maria Vanina Martinez, Ricardo O. Rodríguez, and Gerardo I. Simari, eds.

Volume 38
Logic, Intelligence and Artifices. Tributes to Tarcísio H. C. Pequeno
Jean-Yves Béziau, Francicleber Ferreira, Ana Teresa Martins and Marcelino Pequeno, eds.

Volume 39
Word Recognition, Morphology and Lexical Reading. Essays in Honour of Cristina Burani
Simone Sulpizio, Laura Barca, Silvia Primativo and Lisa S. Arduino, eds

Tributes Series Editor
Dov Gabbay   dov.gabbay@kcl.ac.uk

Word Recognition, Morphology and Lexical Reading

Essays in Honour of Cristina Burani

edited by

Simone Sulpizio

Laura Barca

Silvia Primativo

Lisa S. Arduino

ISBN 978-1-84890-304-3

College Publications
Scientific Director: Dov Gabbay
Managing Director: Jane Spurr

http://www.collegepublications.co.uk

Cover design by Laraine Welch

Preface

This volume arises from the scientific meeting held in June 2019 to honor Cristina Burani on her formal retirement from the Institute of Cognitive Sciences and Technologies (Italian National Research Council). The program dealt with the main topics of Cristina Burani's scientific career, central to Italian Psycholinguistics. The volume contains the contribution of the speakers, which are colleagues and friends of Cristina.

Table of Content

Part III – Neuropsychology

Part IV – Morphology

Introduction

This volume is dedicated to the career of Cristina Burani.

On the occasion of her retirement we, just a handful of the students she has trained and grew up, have decided to organize a meeting to honor her scientific career. To do it the best we can – i.e., in Cristina's style – we bring together many of her colleagues and friends, academics and others, with the common intention of celebrating Cristina.

Below we trace only some of the most relevant lines of Cristina's career, intimately intertwined with her life.

Cristina is indisputably a key scholar for Italian Psycholinguistics.

Her studies have focused on language and lexical processes since the dawn of her scientific activity.
After an initial period dedicated to the study of text comprehension and its educational implications, with the beginning of her career at the CNR (since 1982) Cristina's work has focused on the study of the mechanisms underlying lexical processes in adults with and without acquired neurological disorders, as well as children with typical development and children with developmental dyslexia. Her studies on developmental dyslexia have also contributed to the creation of RIDInet, a recent platform for on-line rehabilitation of Learning Disabilities.

Although the investigation of lexical processing goes throughout Cristina's research activity, she has always been able to tackle the research challenges with a multifaceted approach that winked to neuropsychology, developmental psychology and linguistics. This is also evidenced by the high number of national and international scientific collaborations she developed along her career, as well as the countless opportunities for discussion she organized, in the form of workshops and seminars, which stimulated collaboration and debate between European researchers.

Her activity often took her abroad, as a visiting researcher at some prestigious institute for the study of mind and brain (e.g., Johns Hopkins University (Baltimore, USA); Max-Planck-Institut für Psycholinguistik (Nijmegen, the Netherlands); Center for Biomedical Functioning Imaging (University of Aberdeen, UK); she also took active part in numerous projects funded by national and international organizations (e.g., Research Council, Canada; European Union; International Dyslexia Association, USA).

Crisitina's career was carried out continuously within the CNR, tacking on increasingly significant scientific roles, concluding her career in 2018 as Research Director (the highest level in the hierarchy of CNR scholars). She had great care in fostering the relationship between the CNR and the University, creating lively research programs and training exchanges.

She committed to ensuring that the CNR was not just a research centre, but that it also became a reference point for scientific training. Some proofs of her success in this mission are: the first joint doctoral scholarships between the CNR and the University (Lisa S. Arduino, Laura Barca, Stefania Marcolini), the scientific training of Early stage (Despina Paizi) and Experienced Researchers (Maximiliano A. Wilson) within the Marie Curie Research Training Network 'Language and Brain' (RTN:LAB), and networking program NETWORDS – European Science Foundation (ESF). She has also co-supervised many undergraduate students (including two of the Editors of this volume, Silvia Primativo and Simone Sulpizio), an activity that she has conducted with precision, patience and passion, which are Cristina's trademark.

Strongly believing in cooperation, and having a strong sense of institutions, Cristina has also actively worked at the birth of the Italian Association of Psychology (AIP); She is among the Italian scholars who founded the AIP in 1992,, and its first Honorary Secretary.

We, editors of this volume, will always be indebted to Cristina for the passion and commitment that she has put into our training. Her always constructive critics, tremendous scientific knowledge, passion for work and kindness made her an irreplaceable mentor and collaborator to all of us.
Cristina has not only transmitted us the passion for research, but also the scrupulousness and fairness towards the collaborators. She always encouraged our strengths and played gently of our weaknesses. She has been really demanding but she always found the right words when necessary.
Thanks Cristina, we hope you will finally enjoy your numerous extra-scientific passions, which contribute to make you not only a great scientist but also a fantastic human being.

With endless esteem and affection,

Lisa S. Arduino, Laura Barca, Silvia Primativo, and Simone Sulpizio

Orthography and Reading: Thoughts From Greyfriars Bobby

Andrew W. Ellis[1]*

[1] Emeritus Professor of Psychology, University of York, UK

*andy.ellis@york.ac.uk

Abstract.

After a brief discussion of effects of spelling-sound regularity or consistency on reading in English, I review the work Cristina Burani, Laura Barca and I did on 'rule complexity' effects in Italian word naming, demonstrating that words containing complex spelling-sound correspondence rules involving the letters *c* and *g* are read aloud more slowly than words containing only simple rules, particularly if the words are of lower frequency. The presence of frequency and lexicality effects in Italian show that Italian readers create lexical representations which play a part in identifying and pronouncing familiar written words. A growing body of evidence suggests, however, that skilled readers of transparent orthographies like Italian and Spanish habitually make more use of low-level, sublexical correspondences than do readers of English (a notoriously opaque orthography). I review that evidence, including studies of word learning in English and Spanish which show that Spanish word naming demonstrates larger effects of letter length (an indicator of the involvement of low-level correspondences) than English word naming. Also, Spanish readers switch more slowly and less completely than English readers to whole-word reading and parallel processing of letters in words as unfamiliar words become familiar. I end with some words of thanks to Cristina Burani for encouraging my interest in reading across different orthographies and for being an excellent friend and collaborator.

I first met Cristina Burani at a workshop on The Cognitive Neuropsychology of Language in Venice in 1985. Our research collaboration began, however, over lunch in an Edinburgh pub called the Greyfriars Bobby Bar. That was in September 2001 during a conference of the European Society for Cognitive Psychology.

Cristina and I were both interested in how differences in the characteristics of written language (*orthography*) might affect the way people read words. We knew (of course) that Italian has a logical, reliable, transparent orthography which means that when a reader encounters a new word, they have a high degree of confidence in how it should be pronounced. We also knew that the situation is very different in English where the orthography is so untrustworthy that it is considered 'opaque'. Cristina and I only had soft drinks in Greyfriars Bobby Bar, but if we had fancied beer, we might have ordered two *pints*. The word *pint* is often used as a prime example of an irregular or exceptional word in English, contrasting as it does with *hint, mint, sprint,* etc. Someone attempting to read *pint* for the first time is very likely to mispronounce it.

It is not just rogue words like *pint* that create problems in English. When attempting to read *dough* for the first time, should you rhyme it with *cough, rough, plough, though, borough* or *through*? Each of those words ends in *-ough* but each one has a different pronunciation. (The answer is *though.*) Should *shear* rhyme with *bear, fear* or *year,* or *brow* with *cow* or *show*? The problem is compounded by the fact that many of the commonest words have irregular pronunciations or come from inconsistent word families (e.g., *come, home, four, hour, gave, have, gone, one, phone*).

Cristina and I knew that skilled readers of English are slower to read aloud irregular or inconsistent words than words with regular, consistent spellings (e.g., words ending *-ail* like *fail, hail, snail* and *trail*). We also knew that the difference in reading latencies to irregular/inconsistent and regular/consistent words in English is greater for relatively uncommon, low frequency words than it is for common, high frequency words (see Monaghan & Ellis, 2002, for a review). The usual assumption is that when English words are read for the first time, that is done using low-level letter-sound ('grapheme-phoneme') correspondences. As words become familiar, whole word representations are created that contribute to their pronunciation. If a word is of low frequency, it continues to be read by a combination of lexical and sublexical correspondences ('mapping') on the few occasions it is encountered. If it is a regular, consistent word, those high- and low-level correspondences converge on the same pronunciation which will be produced rapidly and with ease. If it is an irregular, exceptional or inconsistent word, the lexical and sublexical correspondences are likely to generate different pronunciations (correct in the case of the lexical mappings; incorrect in the case of the sublexical mappings). The additional processing time required to resolve that conflict is generally held responsible for the slower reading of low frequency irregular / exception words. If the word is of high frequency and encountered on a daily or weekly basis, lexical reading dominates and effects of regularity / consistency are small or non-existent (Monaghan & Ellis, 2002).

The Case of Italian Word Naming

The question Cristina and I debated was whether effects similar (but not identical) to those of regularity / consistency in English might exist for Italian. It is true that written words in Italian can be read aloud with confidence and that Italian does not contain words with exceptional or inconsistent pronunciations, but it is also the case that the rules governing the way individual letters are pronounced are simpler for some letters than others. Most letters in Italian are pronounced the same wherever they occur in a word and whatever letters are around them, but the pronunciation given to the letters *c* and *g* (and sequences containing those letters) varies according to the context in which they occur. Thus, the letter *c* is pronounced as a hard /k/ in most contexts (e.g., *caldo* [hot], *casa* [house]) but as a soft /tʃ/ like the English "ch" when followed by *e* or *i* (e.g., *cento* [hundred], *cinque* [five]. The rules for pronouncing the letter *g* are even more complex and context-dependent.

The question Cristina and I asked ourselves was whether native speakers of Italian might be slower to read aloud words containing context-sensitive rules than words of comparable length and frequency that contain only simple rules. With the help of Laura Barca, we measured naming latencies for skilled readers pronouncing such words (Burani, Barca, & Ellis, 2006). We found that the latencies for words containing only simple letter-sound correspondences (e.g., *bastone* [stick], *tamburo* [drum]) were significantly faster than the latencies for words matched on frequency and length that contained complex, context-sensitive rules (e.g., *compagno* [partner], *granchio* [crab]).

We already knew that high frequency Italian words are read aloud more rapidly than low frequency words (Bates, Burani, D'Amico, & Barca, 2001). We now know that familiar words are read aloud more rapidly than nonwords (Pagliuca, Arduino, Barca, & Burani, 2008). The existence of frequency and lexicality effects in Italian imply that readers of transparent orthographies develop lexical representations for familiar words even though they could, in principle, read all words using low-level correspondences. A second experiment in Burani et al. (2006) compared the effect of rule complexity in low and high frequency Italian words and found that effect to be greater for low than high frequency words (like regularity / consistency effects in English). A follow-up study found an effect of rule complexity in Italian children aged 8 to 10 years and a similar interaction between complexity and word frequency (Barca, Ellis, & Burani, 2007). We proposed that readers of Italian and English develop lexical representations of familiar words but that sublexical processes continue to contribute, particularly when lower frequency words are read. Those processes are evidenced by effects of rule complexity in Italian and regularity / consistency in English. I still believe that conclusion to be valid, but more recent evidence suggests that the unreliability of English orthography pushes readers more completely towards lexical reading while readers of transparent orthographies habitually combine lexical and sublexical reading to a greater extent.

Word Learning in Opaque and Transparent Orthographies

Ziegler and colleagues came up with a terminology that allows us to talk about the contribution to reading of units of different sizes in a way that is relatively theory- and model-free (Ziegler & Goswami, 2005; Ziegler, Perry, Jacobs, & Braun, 2001). They talked about reading using individual letters as the processing units as *small grain reading* and reading on a whole-word basis as *large grain reading*. Importantly for present purposes, Ziegler and colleagues argued that readers of transparent orthographies make more use of small-grain processing than readers of opaque orthographies who favour large-grain processing.

The method Ziegler et al. (2001) used to assess that theory was to examine the effects on naming latencies of the number of letters in a word or nonword. Many studies have reported that letter length exerts a stronger effect on naming latencies for nonwords than familiar words. Reading an unfamiliar word or nonword aloud requires the reader to progress through it from beginning to end converting orthography to phonology using small-grain units. The more letters, the longer that process takes. As a word becomes familiar through repeated exposure, large-grain (lexical) representations are formed that can process the component letters of the word in parallel, reducing the influence of length (see Kwok & Ellis, 2015, for a review). Ziegler et al. (2001) compared the effect of length on reading latencies to words and nonwords in German (a transparent orthography) and English (an opaque orthography). As predicted, they found larger effects of length on word naming latencies in German than English. Interestingly, that difference applied also to nonword naming despite the fact that nonword naming should rely on small-grain processing regardless of the nature of the orthography.

I now turn to studies of word learning which demonstrate the switch from small- to large-grain processing that occurs as words progress from unfamiliar to familiar. Maloney, Risko, O'Malley, and Besner (2009) employed nonwords varying in length from three to six letters as their stimuli. At the start of the experimental session, the nonwords were presented to participants in a random order with instructions to read each one aloud as quickly as possible. The nonwords were then repeated in a different order, again with instructions to read each one aloud as quickly as possible. This was continued until the nonwords had been read aloud four times. Maloney et al. observed that latencies to the nonwords reduced across presentation blocks and that the large effect of length for first presentations became much smaller by the fourth block. They argued that faster naming latencies and a reduction in the length effect were the hallmarks of word learning and that relatively few presentations were required for skilled readers to create lexical representations of novel words.

Rosa Kwok and I developed and extended that method. I n Kwok and Ellis (2015), we used sets of 8 four-letter, one syllable nonwords (e.g., *brup, carg*) and seven-letter, two-syllable nonwords (e.g., *blispod, drentcy*). As in Maloney et al. (2009), the nonwords were randomly interleaved and presented to skilled, adult readers who were instructed to read each one aloud as quickly as possible. The mean naming latencies in block one were 588 ms. for the short nonwords and 688 ms. for the longer ones. The nonwords were then presented for a nine more blocks, in a different random order each time. Naming latencies reduced across the first five or six blocks before asymptoting at around 480 ms. for the short nonwords and 490 ms. for the longer ones, a difference that was no longer significant. That pattern was replicated in two further experiments using different participants and different stimuli. Experiment 3 brought the participants back into the lab after seven days and found good retention of learning across that interval. By block four of day 7, the difference in latencies to short and long nonwords was down to 1 ms., suggesting that lexical representations were operating and processing was now fully parallel.

Kwok, Cuetos, Avdyli, and Ellis (2017) used the same method to compare word learning in English and Spanish (another transparent orthography). This time, however, the 4- and 7-letter nonwords were randomly interleaved with 4- and 7-letter high and low frequency words. English and Spanish versions of the experiment were run with skilled readers in York, England, and Oviedo, Spain. There were ten blocks of trials on day one with ten more 28 days later.

English naming latencies were faster overall for words than nonwords though no overall frequency effect was observed. As in previous experiments, latencies to English nonwords decreased across blocks on day one and the difference between reaction times to short and long nonwords diminished. Clear benefits of word learning were detectable after the 28-day interval, with only a single presentation being required to bring naming latencies for the English nonwords down to the level they reached at the end of the first session. Taken together, the results of this and other studies suggest that around five to eight encounters with a novel word are sufficient for skilled readers of English to establish robust lexical representations and achieve a more-or-less complete transition from small-grain to large-grain processing. (Adult dyslexic readers require more exposures; see Kwok & Ellis, 2014.)

The results of the Spanish experiment were rather different. Naming latencies were faster to words than nonwords and faster to high than low frequency words. Echoing Ziegler et al.'s (2001) results for German, there were length effects for both words and nonwords in Spanish. Length affected low more than high frequency words but there were effects for both types of word. On day 1, the effect of length on latencies for Spanish nonwords was similar in magnitude to that seen for English nonwords but the effect persisted longer across blocks. Length effects also persisted longer on day 28 for Spanish than for English nonwords. There was good retention of Spanish learning the 28-day interval. Kwok et al. (2017) proposed that while the orthographic properties of English favour a relatively complete switch from small- to large-grain processing as a new word becomes familiar, reading in Spanish habitually uses a combination of small- and large-grain processing, with the contribution of small-grain processing to reading familiar words being detectable in continuing length effects for high- and (particularly) for low-frequency words.

Naming latencies in Kwok et al. (2017) were noticeably slower in Spanish than English for both words and nonwords. The authors speculated that the habitual use of 'mixed grain' reading in Spanish, involving both lexical and sublexical correspondences, may result in slower overall naming than in English where large-grain reading predominates. A similar conclusion was reached by Marinelli, Romani, Burani, McGowan, & Zoccolotti (2016) in a comparison of word and nonword naming by English and Italian children aged 7 to 11 years. The stimuli were high frequency words, low frequency words and nonwords varying in length. Words and nonwords were presented in separate blocks of trials. Both language groups showed faster naming of words than nonwords and faster naming of high than low frequency words. The impact of length on naming latencies was greatest for nonwords, intermediate for low frequency words and smallest for high frequency words, particularly in the older children. That pattern was more marked for the English readers than for the Italian readers. Marinelli et al. concluded that reading in both Italian and English becomes progressively more reliant on larger processing units with age, but that reliance on lexical units occurs earlier in English than Italian, resulting in a stronger effect of lexical variables and a more marked reduction in the effect of length effect as children grow older. In Italian, reliance on larger units occurs later and is less marked.

As in Kwok et al. (2017), naming latencies were longer in the transparent language (Italian) than in English. Marinelli et al. (2016) proposed that the difference may be due in part to the fact that English readers rely more exclusively on larger processing units and a more parallel processing mode where Italian readers routinely combine lexical with sublexical processing when generating the pronunciation of a written word. If so, that difference between languages and orthographies extends to nonword read for the first time as well as to reading familiar words.

This mini-review comes full circle if we ask whether lexical representations play an equal part in the processing of all Italian words or whether the orthographic characteristics of words influence the degree to which large-grain units control processing. Intuitively, it seems possible that the difficulty of converting complex words like *figlio* [son], *cuscino* [pillow] and *giaguaro* [jaguar] from orthography to phonology on a sublexical basis might tip processing more towards large-grain units while the relative simplicity of *trota* [trout], *fratello* [brother] and *sapone* [soap] might encourage mixed-grain processing. If so, then length effects should be larger for simple than complex words in Italian. That prediction could be hard to test using real-word stimuli. It might be easier to generate nonwords of different lengths containing complex rules or only simple rules then present them repeatedly to Italian readers for rapid naming. If the prediction is correct, latencies to complex words of different lengths should converge more than latencies to simple words where the length effect should continue longer, even indefinitely.

***

What began as a shared lunch in Greyfriars Bobby Bar developed into a lasting friendship and collaboration during which I came to appreciate Cristina's careful, meticulous approach to research and her ability to worry away at a problem until she is convinced that she has a plausible and defensible answer. Thank you Cristina, and enjoy your retirement.

References

Bates, E., Burani, C., D'Amico, S., & Barca, L. (2001). Word reading and picture naming in Italian. *Memory & Cognition, 29,* 986-999.

Burani, C., Barca, L., & Ellis, A. W. (2006). Orthographic complexity and word naming in Italian: Some words are more transparent than others. *Psychological Bulletin and Review, 13,* 346-352.

Barca, L., Ellis, A. W., & Burani, C. (2007). Context-sensitive rules and word naming in Italian children. *Reading and Writing, 20,* 495-509.

Kwok, R. K. W., & Ellis, A. W. (2014). Visual word learning in adults with dyslexia. *Frontiers in Human Neuroscience, 8,* 1-12.

Kwok, R. K. W., & Ellis, A. W. (2015). Visual word learning in skilled readers of English. *Quarterly Journal of Experimental Psychology, 68,* 326-349.

Kwok, R. K. W., Cuetos, F., Avdyli, R., & Ellis, A. W. (2017). Reading and lexicalisation in opaque and transparent orthographies: Word naming and word learning in English and Spanish. *Quarterly Journal of Experimental Psychology, 70,* 2105-2129.

Maloney, E., Risko, E. F., O'Malley, S., & Besner, D. (2009). Tracking the transition from sublexical to lexical processing: On the creation of orthographic and phonological lexical representations. *Quarterly Journal of Experimental Psychology, 62,* 858-867.

Marinelli, C. V., Romani, C., Burani, C., McGowan, V. A., & Zoccolotti, P. (2016). Costs and benefits of orthographic consistency in reading: evidence from a cross-linguistic comparison. *PLoS ONE, 11*: e0157457.

Monaghan, J., & Ellis, A. W. (2002). What, exactly, interacts with spelling-sound consistency in word naming? *Journal of Experimental Psychology: Learning, Memory & Cognition, 28*, 183-206.

Pagliuca, G., Arduino, L. S., Barca, L., & Burani, C. (2008). Fully transparent orthography, yet lexical reading aloud: The lexicality effect in Italian. *Language and Cognitive Processes, 23*, 422-433.

Ziegler, J., & Goswami, U. (2005). Reading acquisition, developmental dyslexia, and skilled reading across languages: A psycholinguistic Grain Size Theory. *Psychological Bulletin, 131*, 3-29.

Ziegler, J., Perry, C., Jacobs, A., & Braun, M. (2001). Identical words are read differently in different languages. *Psychological Science, 2*, 379-384.

Some reflections on lexical stress-related effects in word recognition and reading aloud

Lucia Colombo[1]*

[1] Dipartimento di Psicologia Generale, Università di Padova

*lucia.colombo@unipd.it

Abstract

In this short review I focussed on some aspects of the processing of lexical stress assignment in reading and word recognition, considering in particular the work that has been done in Italian, by Cristina Burani and her collaborators and by myself.

I got interested in lexical stress just before the '90s by observing the strong predominance of a stress type over the others in the Italian language, and thinking that this aspect might be analogous to the regularity in the grapheme-phoneme correspondence so much investigated in those years in the literature on cognitive process involved in reading and in the neuropsychology of reading impairments. I was mainly attracted by the idea that people learn distributional properties of their language and apply this learning to the task at hand. Accordingly, in my (Colombo, 1992) paper I pointed out three potential sources of learning about the distribution of stress patterns in the Italian language: 1) the relative distributions of the three stress patterns overall in the language (penultimate syllable stress, antepenultimate syllable stress, and last syllable stress, in order of numerosity of types); 2) the distribution of stress patterns associated to specific orthographic units (stress neighbours), substantially formed by the VCV rhyme of the word: nucleus of penultimate syllable plus last syllable, like -one in pitone (python). In my view, these sources of information had different importance, and effects, depending on the task. These types of information would be used in combination, or in alternative to 3) the lexical information acquired with knowledge of the vocabulary.

To my knowledge, the interest of Cristina Burani for lexical stress started with her paper in collaboration with Alessandro Laudanna, in which lexical stress was proposed to have a role in the planning of articulation, during phonological encoding (Laudanna, Burani, Cermele & Parisi, 1989). Progressively, a substantial body of experimental data about Italian lexical stress processing accrued. Burani and Arduino (2004) and Burani, Paizi, Sulpizio (2014) found that the consistency of neighbours was a more important factor than the overall stress distribution in accounting for both word reading aloud and lexical decision. Similarly, Colombo and Zevin (2009) investigated an important issue related to the assignment of stress during the process of reading aloud, with a continuous priming paradigm. The issue was whether there is an abstract stress frame (a rhythmic frame), independent of segmental information, and whether this frame can be primed. In this paradigm words were preceded by five nonwords primes in a continuous list (i.e., word target and relative nonwords primes were not in a single trial, separated by other trials, each stimulus was part of a sequence of continuous trials). This procedure was thought to investigate priming in a more "natural" way, compared to the usual procedure in which prime and target form a single trial. The results showed evidence for stress priming, in particular, regularization errors and irregularization errors, but only when the nonword context induced sub-lexical reading. Moreover, the data did not show any strong tendency to apply the dominant (penultimate syllable) stress across the board, confirming Burani and Arduino's (2004) claim.

Abstract representation of stress?

The issue of an abstract representation of stress was important in order to model the possible ways in which stress is assigned, and precisely at what stage of the reading process. It was noted that in reading aloud not only the process of accessing the phonological representation of a word/nonword is involved, but also its actual production, thereby including also the processes of phonological encoding and articulation. Reference to one of the most popular models of word production (Levelt, 1989; Levelt, Roelofs & Meyer 1999) for example, suggested that syllabic frame in which the phonological segments had to be inserted, and metrical frame, indicating the number of syllables and position of tonic syllable, were independently processed. Colombo and Zevin's (2009) results agreed with this view, but also showed that stress priming was only possible when reading was sub-lexically driven (i.e., a nonword context). Differently, Sulpizio, Job and Burani (2012b) found that target words primed by stress congruent words were read faster than when primed by stress incongruent primes (see also, Sulpizio et al., 2012a). In both papers it was concluded that stress can be represented as abstract from segmental information. The implication would be that one way to assign stress in reading aloud is by combining segmental information from the activated lexical representations, or from the spelling-sound conversion procedure, with information about number of syllables and position of tonic syllable indicated by the metrical frame. This notion has been used also in a computational model of Italian word reading in which lexical stress was represented (Perry, Ziegler & Zorzi, 2014). When the context in which processing occurs is a word (both prime and targets are words) the metrical frame would be suggested by the prior trial (may be with a higher activation level), and if the latter is congruent with that of the ongoing trial, responses would be fast, otherwise there would be a delay. This would occur during the stage of phonological encoding which is presumed to be involved in reading aloud (i.e., tasks involving production; Sulpizio, Arduino, Paizi & Burani. 2013; Sulpizio, Spinelli & Burani, 2015).

This interpretation is indirectly supported by a recent study by Protopapas et al., (2016; see also Protopapas, Gerakaki, & Alexandri, 2006) who reported data in Greek, showing no priming for words or pseudowords sharing the first syllable with the prime, and congruent or incongruent with the stress pattern of the prime. The lack of priming was found both within- and across modalities, in lexical decision. Given that the first syllable of prime and target overlapped, presumably the target was among the neighbours activated by the prime. If an abstract stress representation is used in word recognition, it should have pre-activated the target with a congruent stress pattern. But as no priming effect was found, the conclusion was that stress priming effects can only be found in reading aloud, adding further support to the view that they are presumably located within the production stage, (i.e., the phonological encoding stage).

Summarizing, these studies have shown that there is an abstract representation of stress, but this abstract representation is probably used only in tasks involving production, for example, in reading aloud. In order to further explore this idea, Sulpizio et al. (2013; 2015) proposed that in order to start articulation the position of the tonic syllable must be specified. When the tonic syllable is the initial syllable, articulation can start immediately, while when the tonic syllable is further on along the word, articulation would be delayed. This account implies a serial mechanism, and predicts that words with stress on the antepenultimate syllable should be named faster than words with penultimate syllable stress.

That the motor program for articulation works serially from left to right is plausible, given that the spoken output is temporally spread in time. Indeed, Sulpizio and Job (2015) investigated this idea further in a number of experiments in which target words to be read aloud were preceded by different primes. In their Experiment 3, prime and target shared the first syllable and did (FEgato- FEcola; starch-liver) or did not (feNIce- FEcola; starch-phoenix) share the stress position. Although the prime was masked and not visible, there was a supra-segmental effect on the latencies to read the target, which was different depending on whether the target had penultimate or antepenultimate stress. In the latter condition (FEgato -FEcola) latencies were shorter in the congruent compared to both incongruent (feNIce-FEcola) and control (a sequence of %) conditions. However, for the penultimate stress targets there was no effect of congruency: both congruent and incongruent conditions were faster than the control condition. Apparently, when the first syllable was not stressed, participants could programme the first unstressed syllable, on the basis of the segments, and postpone the decision on which of the following was the stressed syllable. Thus it was irrelevant whether primes were or were not congruent. Note that supra-segmental priming is only obtained when the task is reading aloud, not with lexical decision. As mentioned above, when the task is lexical decision no priming was obtained by Protopapas et al. (2016) in Greek. Thus, Sulpizio and Job's study supports the hypothesis that these effects are to be located during the stage of phonological encoding, which is obligatory when the target must be produced.

Sources of information for lexical stress and statistical bias

In ordinary situations words are not primed by other words, so what the data from these priming experiments tell us is that there is an abstract representation of stress, which can be used during phonological encoding, but is unlikely to be used in word recognition. If we accept that stress has an abstract representation and can be assigned independently of segmental information by processes specific to a task, the next question is: how is the position of the tonic syllable decided in reading aloud, on which basis? And under which conditions is this abstract representation used? As Italian stress is not fixed, lexical information is the most secure source of information about the position of the stressed syllable. However, also novel words must receive stress, and this information is not available for them. Moreover, the fact that many studies showed effects of different types of stress pattern distributions (stress neighbourhood, overall distribution) suggests that this type of information is accessed before lexical access occurs, at least for words that are not very frequent.

One possibility discussed in the literature, but that has received weak support, is that in the absence of other indications, like for example a previous presented word/nonword, or of orthographic cues, the most plausible candidate is the most frequent stress pattern in the language. If we believe that distributional information about the language characteristics, implicitly learned, is important, as shown by the effects of stress neighbourhood, then why shouldn't the general distribution of stress pattern be influential under some form? But under what form? A proposal made by Levelt and coll. is that there is a default mechanism, conceptualized as a kind of rule, that in the absence of other information selects the stress pattern most frequent in the language. Colombo (1992, page 987) made a suggestion only apparently similar, that implicit knowledge about the dominant pattern in the language (Italian) "should provide a bias in favour of the frequent stress pattern that may influence the retrieval or computation of the pronunciation and it can be conceived as a kind of rule ". This (bias) might be used as a default, where default should be intended as "a selection made usually automatically or without active consideration due to lack of a viable alternative" or of a specification (Merriam-Webster Dictionary).

As mentioned above, several studies that have manipulated the type of stress (penultimate syllable stress, which is the dominant pattern (about 80% of all types, counting all word types, independently of the number of syllables), and antepenul-timate syllable, which is much less frequent (about 20%) together with type of neighbourhood (friends vs enemies) have shown that endings are much stronger cues to stress, and are more likely to be used by participants in a reading aloud task. In reading aloud, beginning readers show This bias towards the dominant penul-timate stress is apparently shown by early readers (Colombo, Deguchi, & Boureux, 2014; Sulpizio & Colombo, 2013), while with the increase in reading skills, readers rely on a word- specific source of information, stress neighborhood (Burani & Ardu-ino, 2004; Burani, Paizi, & Sulpizio, 2014; Sulpizio, Arduino, Paizi, & Burani, 2013).

In these studies, the sources of information for stress are described as biases that the reading system obtains from the implicit knowledge acquired with lan-guage experience. Reference to a "rule" is intended metaphorically, not as the claim of the use of precise algorithms. In contrast, Rastle & Coltheart (2000) proposed a model of stress assignment for the English language based on a mechanism apply-ing rules. However, two recent studies have provided very little evidence in favour of a rule mechanism operating to assign stress (Ktori, Mousikou & Rastle, 2018; Mousikou et al., 2017). Mousikou et al., (2017) employed a very large nonword database in order to find which are the possible cues to stress in English. They analysed the nonword pronunciations produced by a sample of participants and found several types of orthographic cues to stress, in particular, prefixes, vowel length and orthographic complexity. Ktori et al. (2018) manipulated these most important types of sublexical cues with the aim to explore how they influence the assignment of second syllable stress in reading aloud, given that the most frequent stress pattern in English is on the first syllable. They found that all three types were influent. For example, they found that participants were more likely to assign second-syllable stress when the nonwords had a prefix, and had a long second vowel. Their data showed a clear sensitivity to statistical information related to stress, both in terms of the general distribution of stress, and of orthographic cues to stress.

The fact that statistically-based information, as opposed, for example, to a rule-based approach, is critical in the assignment of stress has received a lot of attention, recently, in the literature. In particular, the role of Bayesian probabilistic inferences in the assignment of stress has been investigated initially by Jouravlev and Lupker (2015) who calculated the probabilities of a stress pattern on the basis of both the general distribution of stress types in Russian (prior probability), and of the evidence from the characteristics of the processed word/nonword (posterior probabilities). Note that in Russian the difference in the distribution of trochaic and iambic patterns is not very strong overall, and becomes large only within the grammatical category of adjectives (80% vs 20%, respectively). The posterior, evidence-based probabilities are grounded on both lexical and sub-lexical information. The former type of evidence (lexical) comes from lexical access, the latter from orthographic information. Jouravlev and Lupker (2015) found that the predictive power of the posterior probabilities was very high, with a strong correlation (r = .94) between the predicted posterior probabilities and the proportion of iambic stress assigned by participants, (independently of whether the words were correctly pronounced or not). Ktori et al. (2018) also applied the Bayesian probabilistic inference approach to the English language and found that posterior probabilities were very good predictors of second stress assignment by participants (r = .49). The implication of this approach is that if there is a dominant stress in a language, it can be used as a baseline, but its influence can be moderated or reversed depending on the strength of the evidence-based probabilities, like orthographic cues, grammatical class, or other sources of information. An important implication is also that in this framework it is not assumed that the assignment of stress depending on the general language distribution is a mechanism different from that assigning stress on the basis of, for example, orthographic cues. Another important implication is that it can easily explain why under some circumstances one type of stress can predominate over the other, independently of whether it is the most frequent stress pattern in the language (e.g., Sulpizio et al., 2013; Spinelli et al., 2016).

Evidence in favour of a bias to apply the dominant stress is not straightforward because most studies manipulated stress neighbourhood as well. It has been shown that endings have a strong role in Italian and if the stimuli are characterized by end-ings biased in one direction (favouring penultimate or antepenultimate stress) the relative stress pattern will emerge, giving rise to contrasting data. Is it possible to find conditions under which no bias is provided by endings? Perhaps when endings are associated to a small neighbourhood or the neighbours are not associated to a single stress pattern, this might be the ideal condition to find effects of the dominant language.

Sulpizio et al., (2013) in a study in which they investigated the factors that de-termine how stress is assigned to nonwords, used strongly biased endings towards penultimate or antepenultimate stress and constructed nonwords with these end-ings. They found that preference for antepenultimate stress for nonwords consist-ent with this pattern was more marked than the preference for penultimate stress, a result which contrasts with the idea that there is a bias for penultimate stress. They also found that number of neighbours was also important, however, such as the ef-fect of consistency between the stress pattern assigned and the corresponding end-ings was partly explained by the greater number of word types in the antepenultim-ate stress neighbourhoods. However, it is not only the ending that drives stress: also the initial part of the nonwords, the type of syllable included in the nonword, the presence of affixes, the words' grammatical class, or the nonword CV structure (Monaghan, Arciuli & Seva, 2016; Spinelli, et al., 2016; Sulpizio et al., 2013; Sulpizio, Spinelli & Burani, 2017) may offer important cues to stress, and their effects should be investigated simultaneously in order to draw firm conclusions.

Dominant stress and word recognition

The role of the dominant stress in word recognition has been investigated in different languages, including English, Spanish, Greek and Russian, in terms of whether word recognition is easier for words with the most frequent stress pattern. In English, Rastle and Coltheart (2000) did not find an advantage for the dominant stress in lexical decision (see also Kelly, Morris, & Verrekia, 1998). Arciuli and Cupples (2006) in contrast found effects of the most typical stress pattern according to the grammatical category, in English: in lexical decision, there were fewer stress errors in nouns with initial stress, and in verbs with second syllable stress, than vice versa.

Protopapas et al., (2016) although they did not find supra-segmental priming in a masked priming lexical decision, found however an overall advantage for the dominant stress in Greek. Jouravlev and Lupker (2014) explored the distribution of stress patterns in Russian: as noted above, only for the grammatical category of adjectives there is an asymmetric frequency distribution, with adjectives stressed on the first syllable more frequent (80%) than those with stress on the second syllable (20%). They found only for this grammatical category an advantage for the dominant stress, which was significant in both latencies (in the subject analysis) and errors.

Recent data with a lexical decision task show that words with penultimate stress are recognized faster and more accurately than words with antepenultimate stress in Italian (Sulpizio & Colombo, 2015; see also Colombo, 1992). In contrast, no effect of stress neighbourhood was found. In Colombo and Sulpizio (2015) words and nonwords were selected in a way that allowed the orthogonal manipulation of type of stress (penultimate and antepenultimate) and stress neighbours (favouring penultimate or antepenultimate stress). One might have expected either a main effect of stress neighbours or an interaction with type of stress, but neither was found. This result seems apparently at odds with data from reading aloud, because in order to make a lexical decision one needs to read the words, and if endings provide strong cues to stress, one might expect that effects of stress neighbours should also emerge in lexical decision.

One possibility might be that effects of stress neighbourhood are very fast-acting, but are overcome by other factors in lexical decision. That is, in reading aloud stress endings provide cues to stress that are exploited at the phonological and phonetic level to construct a phonetic and articulatory code for pronunciation. As lexical decision requires lexical access, endings might activate phonological lexical units, but if access is sufficiently fast, information coming from endings might not be exploited, or their effect might fade away rapidly.

Sulpizio and Colombo (2017) investigated with the same material and paradigm used in Sulpizio and Colombo (2015) the temporal course of stress-related effects through ERPs. They found that the advantage of penultimate stress words appears relatively late, while within about 150 ms from word onset an effect of stress endings appeared. In particular, penultimate stress words with many stress enemies (e.g., seNIle, senile, whose ending –ile is most frequently occurring in antepenultimate stress words, like FAcile) displayed a larger posterior P1 and a more anterior N1 complex compared to both penultimate stress words (with many stress friends) and to antepenultimate stress inconsistent words, which did not differ. Apparently, the first components reflect orthographic processing required to isolate the ending, independently of whether the word is phonologically consistent or inconsistent with its ending and of the word's stress pattern. For example, the ending -ita is typical of a dominant stress neighbourhood and the word graNIta is consistent because it has penultimate stress, while BIbita, with antepenultimate stress, is inconsistent with its ending. Apparently, the very first stage of processing reflects the activation pattern of stress neighbours, independently of the stress pattern of the word in which they are included. Thus the endings would be used to isolate orthographic units in the word string and would emerge because of their redundancy in Italian. Further, the N1 component might reflect attentional mechanisms that look for potential useful cues in the string.

Following these early components, a later component was also observed, differentiating words that are consistent (*graNIta- MIssile*) but have a different stress pattern. This later positive component has been interpreted as a marker of lexical stress, which would be used to access the lexical phonological representation. Summarizing, the ERPs results suggest that in word recognition there is an early activation of endings, partially driven perhaps by attentional mechanisms, that provide cues to stress. Later, following lexical activation, the two typical stress patterns are differentiated. The fact that the behavioural data of the lexical decision task do not show stress neighbourhood effect would be explained therefore by the fact that they are elicited very early, followed by lexical activation of the word, when they are no longer useful. In reading aloud, in contrast, the activation of endings can be used, besides for cueing the probable stress of the word, to drive articulation, as suggested by Sulpizio et al., (2013) and Spinelli et al., (2016) therefore they would persist for longer.

A further issue that deserves reflection is how the dominant stress produces an advantage in word recognition. It is clear that orthographic cues to stress provided by endings can only be used in a probabilistic way: an ending is associated to a stress pattern, but does not incontrovertibly indicate the stress pattern of the word in which it is included. However, the correct stress pattern is only available through lexical consultation. Therefore, the advantage of dominant stress in lexical decision might be due to easier access to the phonological lexicon for words with dominant stress.

Alternatively, the fact that penultimate stress words are overwhelmingly more numerous might act as a bias to expect a dominant stress and therefore to facilitate responses consistent with the bias.

In a Bayesian framework (Norris, 2013; Jouravlev & Lupker, 2015; Ktori et al., 2018) a "word" response in lexical decision is given when the participant is confident that the input is a word. As noted, this confidence is gathered collecting evidence from prior probabilities (e.g., word frequency and knowledge of the dominant stress in the language) and posterior probabilities from the orthographic input (e.g., endings and their neighbourhoods or other sorts of cues). However, evidence from endings adds much noise to the system because it is in competition: many endings cue one type of stress, but in contrast with the stress pattern of the word itself, as is the case for words with many stress enemies. Therefore, the influence of this information is less useful in lexical decision. In contrast, information about the dominant language provides a high validity cue in Italian, given the relative proportion of the stress patterns. This might explain why in Sulpizio and Colombo (2017) the effects of stress neighbours are found very early, but their influence fades away and the responses reflect the effect of the overall distribution of stress. Moreover, as Norris (2013) noted, a Bayesian framework naturally explains differences among tasks, like the different influence of stress neighbours in reading aloud and lexical decision.

Conclusions

In conclusion, almost twenty years of research focussed on how lexical stress is assigned in reading aloud and lexical decision have produced a fair amount of research, data, and theoretical discussions, many of which carried out by Cristina Burani and her collaborators. These studies have stimulated the investigation of lexical stress in reading aloud and lexical decision in other languages, and the production of computational models (Pagliuca & Monaghan, 2010; Perry, Ziegler & Zorzi, 2010; 2014; Rastle & Coltheart, 2000; Seva, Monaghan, & Arciuli, 2009) that have allowed simulations and comparisons with real data. Thus, they have provided a start up for thinking of an issue that before had been quite neglected in the dominant literature on word recognition and reading.

References

Arciuli, J., & Cupples, L. (2006). The processing of lexical stress during visual word recognition: Typicality effects and orthographic correlates. *The Quarterly Journal of Experimental Psychology, 59,* 920-948.

Burani, C., & Arduino, L. S. (2004). Stress regularity or consistency? Reading aloud Italian polysyllables with different stress patterns. *Brain and Language, 90,* 318-325.

Burani, C., Paizi, D., & Sulpizio, S. (2014). Stress assignment in reading Italian: Friendship outweighs dominance. *Memory & Cognition, 42,* 662-675.

Colombo, L. (1992). Lexical stress effect and its interaction with frequency in word pronunciation. *Journal of Experimental Psychology: Human Perception and Performance, 18,* 987.

Colombo, L., & Sulpizio, S. (2015). When orthography is not enough: The effect of lexical stress in lexical decision. *Memory & cognition, 43,* 811-824.

Colombo, L., & Zevin, J. (2009). Stress priming in reading and the selective modulation of lexical and sub-lexical pathways. *PloS one, 4:*e7219.

Colombo, L., Deguchi, C., & Boureux, M. (2014). Stress priming and statistical learning in Italian nonword reading: Evidence from children. *Reading and Writing, 27,* 923-943.

Jouravlev, O., & Lupker, S. J. (2014). Stress consistency and stress regularity effects in Russian. *Language, Cognition and Neuroscience, 29,* 605-619.

Jouravlev, O., & Lupker, S. J. (2015). Lexical stress assignment as a problem of probabilistic inference. *Psychonomic Bulletin & Review, 22,* 1174-1192.

Kelly, M. H., Morris, J., & Verrekia, L. (1998). Orthographic cues to lexical stress: Effects on naming and lexical decision. *Memory & Cognition, 26,* 822-832.

Ktori, M., Mousikou, P., & Rastle, K. (2018). Cues to stress assignment in reading aloud. *Journal of Experimental Psychology: General, 147,* 36.

Laudanna, A., Burani, C., Cermele, A., & Parisi, D. (1989). Effetti sillabici e accentuali nella lettura di parole nuove [Syllabic and stress effects in reading new words aloud]. *Giornale Italiano di Psicologia, 16,* 119-140.

Levelt, W. J. (1993). *Speaking: From intention to articulation*(Vol. 1). MIT press.

Levelt, W. J., Roelofs, A., & Meyer, A. S. (1999). A theory of lexical access in speech production. *Behavioral and brain sciences, 22,* 1-38.

Monaghan, P., Arciuli, J., & Seva, N. (2016). Cross-linguistic evidence for probabilistic orthographic cues to lexical stress. *Linguistic rhythm and literacy,* 215-236.

Mousikou, P., Sadat, J., Lucas, R., & Rastle, K. (2017). Moving beyond the monosyllable in models of skilled reading: Mega-study of disyllabic nonword reading. *Journal of Memory and Language, 93,* 169-192.

Norris, D. (2013). Models of visual word recognition. *Trends in cognitive sciences, 17,* 517-524.

Pagliuca, G., & Monaghan, P. (2010). Discovering large grain sizes in a transparent orthography: Insights from a connectionist model of Italian word naming. *European Journal of Cognitive Psychology, 22,* 813-835.

Perry, C., Ziegler, J. C., & Zorzi, M. (2010). Beyond single syllables: Large-scale modeling of reading aloud with the Connectionist Dual Process (CDP++) model. *Cognitive psychology, 61,* 106-151.

Perry, C., Ziegler, J. C., & Zorzi, M. (2014). CDP++. Italian: Modelling sublexical and supralexical inconsistency in a shallow orthography. *PloS one, 9:* e94291.

Protopapas, A., Gerakaki, S., & Alexandri, S. (2006). Lexical and default stress assignment in reading Greek. *Journal of Research in Reading, 29,* 418-432.

Protopapas, A., Panagaki, E., Andrikopoulou, A., Gutiérrez Palma, N., & Arvaniti, A. (2016). Priming stress patterns in word recognition. *Journal of Experimental Psychology: Human Perception and Performance, 42,* 1739.

Rastle, K., & Coltheart, M. (2000). Lexical and nonlexical print-to-sound translation of disyllabic words and nonwords. *Journal of Memory and Language, 42,* 342-364.

Ševa, N., Monaghan, P., & Arciuli, J. (2009). Stressing what is important: Orthographic cues and lexical stress assignment. *Journal of Neurolinguistics, 22,* 237-249.

Spinelli, G., Sulpizio, S., Primativo, S., & Burani, C. (2016). Stress in context: Morpho-syntactic properties affect lexical stress assignment in reading aloud. *Frontiers in psychology, 7,* 942.

Sulpizio, S., Arduino, L. S., Paizi, D., & Burani, C. (2013). Stress assignment in reading

Italian polysyllabic pseudowords. *Journal of Experimental Psychology: Learning, Memory, and Cognition, 39,* 51-68.

Sulpizio, S., Boureux, M., Burani, C., Deguchi, C., & Colombo, L. (2012a). Stress assignment in the development of reading aloud: Nonword priming effects on Italian children. In *Proceedings of the Annual Meeting of the Cognitive Science Society* (Vol. 34, No. 34).

Sulpizio, S., Burani, C., & Colombo, L. (2015). The process of stress assignment in reading aloud: Critical issues from studies on Italian. *Scientific Studies of Reading, 19,* 5-20.

Sulpizio, S., & Colombo, L. (2013). Lexical stress, frequency, and stress neighbourhood effects in the early stages of Italian reading development. *The Quarterly Journal of Experimental Psychology, 66,* 2073-2084.

Sulpizio, S., & Colombo, L. (2017). Early markers of lexical stress in visual word recognition. *Memory & cognition, 45,* 1398-1410.

Sulpizio, S., & Job, R. (2015). The segment-to-frame association in word reading: early effects of the interaction between segmental and suprasegmental information. *Frontiers in psychology, 6,* 1612.

Sulpizio, S., & Colombo, L. (2017). Early markers of lexical stress in visual word recognition. *Memory & cognition, 45,* 1398-1410.

Sulpizio, S., Job, R., & Burani, C. (2012). Priming lexical stress in reading Italian aloud. *Language and Cognitive Processes, 27,* 808-820.

Sulpizio, S., Spinelli, G., & Burani, C. (2015). Stress affects articulatory planning in reading aloud. *Journal of Experimental Psychology: Human Perception and Performance, 41,* 453.

Sulpizio, S., Spinelli, G., & Burani, C. (2017). STRESYL. *Written Language & Literacy, 20,* 80-103.

Neural correlates of covert word reading in hearing and deaf adults

Laura Barca[1]*, Antionio Napolitano[2], Marianna Castrataro[1], Pasquale Rinaldi[1], Vittorio Cannatà[2], Maria Cristina. Caselli[1]

[1] Institute of Cognitive Sciences and Technology, ISTC-CNR, Rome. Italy

[2] Pediatric Hospital 'Bambino Gesù' of Rome, Italy.

* laura.barca@istc.cnr.it

Abstract.
The present study compared the neural correlates of covert reading in hearing and deaf adults proficient readers. Participants were asked to covertly read visually presented words or passively attend strings of consonants while being scanned with fMRI. Deaf participants preferentially using spoken Italian (Deaf-SI) and deaf participants preferentially using Italian Sign Language (Deaf-LIS) were also compared. Deaf-LIS showed activation of supramarginal gyrus and precentral regions of the left hemisphere, indicating greater reliance of articulatory component of speech during covert word reading. Compared with hearing participants, Deaf-LIS have higher activations of lingual and fusiform gyri, close to the Visual Word Form Area, suggested to be specialized for printed word recognition. Compared with Deaf-SI participants, Deaf-LIS showed de-activation of posterior cingulate cortex, suggesting greater attentional effort in covert reading. Our results confirm the presence of a cortical network shared by hearing and deaf participants with different communication mode when covertly read words. They also reveal differences in neural activity driven by auditory and linguistic experiences, pointing to a greater role of articulation of speech, lexical-orthographic processing and their corresponding neuronal resources for deaf using sign language.

The visual recognition of printed stimuli typically relies on a left-lateralized network that involves *occipital regions* and *fusiform gyrus* for orthographic processing, then a forward sweep of activation moves into *inferior and middle temporal areas* involved in semantic processing and *perisylvian and inferior prefrontal areas* involved in phonological processing (Barca et al., 2011; Marinkovic et al., 2003; Pammer et al., 2004; Price, 2012). The integrity of the visuo-perceptual sensory system is not mandatory for reading, as congenitally blind individuals can read via touch using Braille and in doing so they activate a neural network strikingly similar to sighted (Reich, Szwed, Cohen, & Amedi, 2011), supporting the 'metamodal theory' of brain function (Pascual-Leone & Hamilton, 2001).

The current study is designed to explore if auditory experience and preferred communication mode (spoken vs signed) affects the neural correlates of written language processing.

Deaf people largely vary in their degree of familiarity with sign language. Deaf people from hearing parents are exposed to spoken language in family and environmental context, and might be less exposed to sing language. Differently, deaf people from deaf signing parents are immediately exposed to a sign language, but are also familiar with the spoken language used in their environment, thus they grow as bilingual. In comparison to hearing people exposed to two spoken languages (i.e., unimodal bilinguals), bimodal bilinguals constitute not one, but many different populations. This variety reflects a wide range of factors such as the time of onset of bilingualism, the amount of exposure to each language and the settings in which each language is acquired and used. In Italy, as in many other countries, only between 5 to 10% of deaf children acquire a sign language natively from deaf signing parents (Caselli, Maragna & Volterra, 2006). Native signers benefit from a more homogeneous language experience than non-native signers, but large individual differences can be found (Rinaldi, Caselli, Onofrio, & Volterra, 2014).

In a previous study we demonstrated that regardless to sign language knowledge, deafness does not necessarily impair the processing of written words (Barca, Pezzulo, Castrataro, Rinaldi, & Caselli, 2013). In Barca et al. study we compared the performances of deaf adults native signers, deaf adults preferentially using spoken Italian, and hearing controls in a visual lexical decision task. The typical lexicality effect, with faster recognition of words over consonant strings (Pagliuca, Arduino, Barca & Burani, 2008; Barca, Bello, Volterra & Burani, 2009), holds only for deaf signers. No such effect was present in deaf participants preferentially using spoken Italian and in hearing controls. We suggested that deaf signers over rely on lexical reading, recognizing the stimuli mainly via '*whole word visual-orthographic processing*'. We speculated that lexical items might benefit from richer visuo-motor or sign-based representations (Barca and Pezzulo, 2017; Morford et al., 2011). At the cerebral level, this might reflect in the recruitment of regions of the brain related to the coding of motor acts and praxis information, such as the inferior parietal lobule (Corina et al., 2007; Pobric, Jefferies & Lambon Ralph, 2010).

In the present study, we employed functional MRI to explore the cortical network recruited by deaf and hearing participants when covertly read lexical stimuli. The purposes of the study were twofold: The first was to shed light on the cortical network involved in covert reading by deaf readers of a shallow orthography (Italian). The second was to test for an effect of communication mode by comparing the performance of deaf participants who preferentially use Italian Sign Language-LIS to communicate (here termed "Deaf-LIS") and those who preferentially use Spoken Italian-SI (here termed "Deaf-SI"), and have less or no competence in LIS.

Although hearing and deaf participants were expected to recruit a similar network of occipitotemporal, temporoparietal, and frontal cortices, differences due to their auditory experience and preferred mode of communication were also expected.

Specifically, we expect between-groups differences at left lateralized pre-frontal sites, typically involved in phonological processing. In the case of *deaf readers*, the uncertainty of auditory input since infancy might prevent the establishing of stable phonological representations of words and letters. Thus, differences might arise at left lateralized structures such as superior temporal areas, supramarginal gyrus, and the opercular part of the inferior frontal gyrus, involved in the building of stimulus pronunciation (see Jobard, Crivello & Tzourio-Mazoyer, 2003 for a review). Moreover, deaf preferentially using sign language might also recruit portions of the motor cortex linked to the programming and execution of hand movements, related to their preferred communication mode. Some activation of left prefrontal regions might be expected in relation to lip-reading and inner speech, however this should be less pronounced compared to hearing and deaf participants using spoken Italian (Aparicio et al., 2007; MacSweeney et al., 2008).

Functional Neuroimaging Experiment

Participants
Participants were 21 university students, expert readers of Italian, who took part in the Barca et al., (2013) study. They were 20-25 years old, and divided in three groups according to their auditory experience and preferred mode of communication. Deaf-LIS were 7 native deaf participants[1] who learned LIS as their

---
[1]

Both groups of deaf participants have sever to profound bilateral sensorineural hearing loss (71 + dB in the better ear).

first language in family context (deaf children from deaf signing parents), and use it daily in different social contexts. Deaf-SI were 7 native deaf participants, born from hearing parents and primarily using spoken language, with limited or no use of LIS. Deaf participants were interviewed before the experiment to obtain detailed characteristics of their deafness, language experience, and their use of hearing aids (see Appendix 1.1 for details). None of the deaf participants had cochlear implant when data were collected.

A group of 7 hearing monolingual participants, native speakers of Italian without any knowledge of sign language was also included in the study. The three groups had comparable cognitive and linguistic skills (see Barca et al., 2013).

Materials and procedure
During the fMRI experiment, participants were visually presented with words and consonant strings. They were asked to attend the stimuli presented and if a word appeared, they had to name it silently 'in their head'; if a string of consonants appeared, they were instructed just to attend the stimulus.

The experimental list comprised 100 Italian words and 100 consonant strings, and is fully described in Barca et al., (2013). All participants were tested individually. Instructions to deaf participants were provided in their preferred communication mode by a hearing research assistant, LIS interpreter and expert in communication with deaf persons.

A block-design was used (10 blocks of 30seconds each, with 20 stimuli per block). The sequence of the events was as follows: first a central fixation cross was presented (400ms), then an experimental stimulus appears visually for 1000ms (ARIAL font, upper case, white print on a grey background), followed by a second central fixation cross. This was used as a catch trial, so that participants had to press a button on the response device held in their left hand if the cross was red (20% of cases).

Data have been acquired in three runs, with a small break in between. Following the acquisition of the functional data, anatomical structure of the brain was acquired. The overall session lasted about 1.5 hours. Details on the fMRI apparatus, image acquisition, processing and analyses are reported in the Appendix 1.2.

Results

Greater activation for covert word reading than passive viewing of consonants strings

---

Results of words vs. consonant strings contrast for each group are reported in Table 1. Overall, Hearings and Deaf-SI participants show a complex pattern of activations, spread bilaterally across occipital, temporal and frontal brain regions. Differently, Deaf-LIS participants present a defined patch of activation strictly lateralized to precentral regions of the left cerebral cortex.

Hearing participants
The contrast elicited significant higher activation for words over consonants strings in regions of the right hemisphere comprising superior and middle temporal cortex (STG and MTG, respectively), superior occipital gyrus (SOG) and left cuneus.

Deaf-SI
Deaf-SI showed enhanced activations in prefrontal regions of the left hemisphere which, together with the anterior cingulate cortex (ACC) and superior/mid temporal gyrus (S/MTC), are thought to be part of a lexical-semantic system (Ardila et al., 2016; Gabrieli et al., 1993). Interestingly, the ACC has been implicated in language switching task (Abutalebi and Green, 2016), and recently suggested to be part of a 'bilingual language control network' (D'Souza and D'Souza, 2016). Corina et al. (2013) report grater ACC activation in word relative to false font task, interpreted as evidence of implicit lexical activation. Within the right hemisphere, activations were found in STG and fusiform gyrus (FG), involved in visual word recognition and lexical semantic processes (Ardila et al., 2016).

Overall, this pattern suggests that Deaf-SI recruit lexical information when covertly read words

Table 1. Location of activation for covert words reading relative to passive viewing of consonants strings, at peaks that were significant at p<0.01 (uncorrected for multiple comparisons).

Approximate location (BAs)	MNI coordinates			
	x	y	z	T value
Hearing				
R superior temporal area (22)	45	4	-13	9.52
R middle temporal area (22)	48	-11	-14	5.04
R superior occipital area (19)	24	-85	26	4.54
Cuneus (VisAssArea - 18)	1	-85	13	4.34
R middle temporal area (21)	48	-15	-14	3.31

Deaf-SI				
L prefrontal cortex (9)	-5	55	25	5.97
L middle temporal gyrus (OutBAs)	-26	-46	6	5.85
R fusiform gyrus (21)	48	-42	-7	5.63
L prefrontal (9)	-12	43	28	5.09
L anterior cingulate gyrus (32)	-8	24	36	4.37
R superior temporal gyrus (22)	48	-11	-10	4.17
L superior and midtemp gyrus (OutBAs)	-40	-50	6	3.32
R fusiform gyrus (37)	38	-73	-13	3.32
Deaf-LIS				
L supramarginal gyrus (OutBAs)	-43	-11	21	3.78
L precentral gyrus (OutBAs)	-43	-7	21	3.58
L precentral gyrus (PrimSensory 1)	-43	-15	12	3.52
L precentral gyrus (PrimSensory 1)	-43	-19	38	3.16

Note. Results are thresholded at p < 0.01 (uncorrected). Coordinates of activations are given in the Montreal Neurological Institute (MNI) space. Broadman areas corresponding to the activations were identified using the Yale MNI to Tal application (Lacadie et al., 2008). VisAssArea = Visual Associate Area; Out BAs = Outside defined BAs.

The activation of left STG is commonly reported in linguistic tasks, far less common is the recruitment of its right homologue (see Ardila, Bernal, Rosselli, 2016). Given the role of the right MTG in semantic processing (Hernandez et al., 2015), retrieval of verbal information (Shallice et al., 1993) and response generation (Mechelli, Friston and Price, 2000), this suggest that hearing participants solve the task with the recruitment of semantic information. Right MTG has also been associated with the processing of auditory information, specifically of voice recognition (Roswandowitz et al., 2018), which might be related to their auditory linguistic experience. Finally, stronger activations for words is also observed in visual association area as superior occipital cortex (SOC) and the cuneus (Ardila et al.,

2016).

The observed right hemisphere activations, thus, might be related to the generation of a verbal response from visual information.

Deaf-LIS

In Deaf-LIS the enhanced activation for words over consonant strings was strictly left lateralized to regions involved in executive control of language production (Ardila et al., 2016). It comprises a locus in the supramarginal gyrus, close to the left primary motor cortex (BA4), which has been associated with phonological processing (Sliwinska et al., 2012), and specifically in tasks involving covert articulation (Pattamadilok et al., 2010). The supramarginal gyrus is included in the dorsal pathway of the reading network, which is specialized for serial letter-by-letter reading (Carreiras, Mechelli, Estévez, & Price, 2007). Moreover, Corina, Lawyer, Hauser & Hirshorn (2013) reported greater bilateral SMG activation for deaf proficient readers during word processing (relative to false fonts) compared to the less-proficient readers during word processing (relative to false fonts).

Additional loci of activations are scattered along the precentral gyrus, close to the portion of the speech-motor cortex consistently reported in overt and covert word production (Barca et al., 2011; Carreiras et al., 2007; Dietz et al., 2005).

Group comparison

In order to evaluate areas differently recruited by the three groups of participants, the words vs consonant strings contrast was evaluated between paired groups' comparison. The comparison leads to a number of significant clusters of activation within the left hemisphere (see Figure 1 and Table 2).

Hearing participants and Deaf-SI

There were no significant differences in the subtraction contrast between Hearing participants and Deaf-SI. Thus, the functional neuroimaging data mirrors the behavioral performance observed in a lexical decision task (Barca et al., 2013).

Hearing participants and Deaf-LIS

Compared to hearing participants, Deaf-LIS showed enhanced activation (of words vs consonants strings) in regions within the occipital gyrus and fusiform gyrus, which are considered to reflect pre-lexical orthographic-phonological mapping (Wheat et al., 2010).

Figure 1. Differences between groups (cover word reading vs passive views of consonant strings)

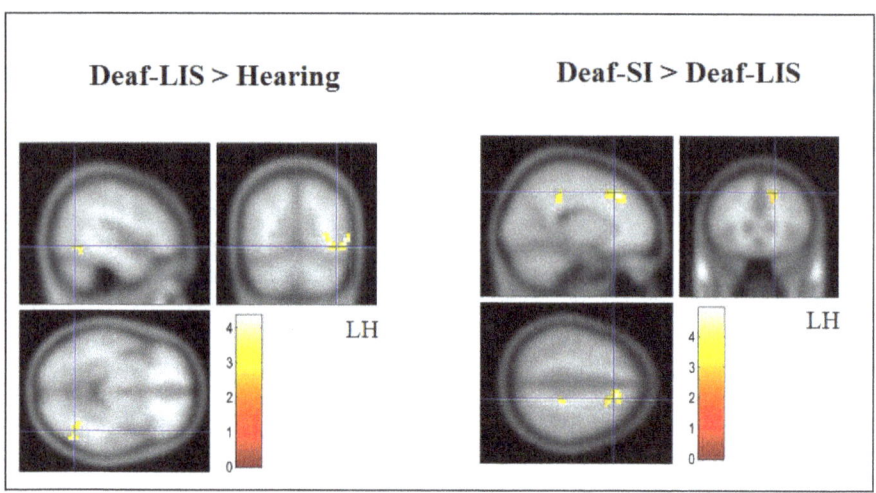

Deaf SI and Deaf LIS
Comparing the two groups of deaf participants, Deaf-SI showed enhanced activation within the posterior cingulate cortex (PCC) (see Table 2). The anterior portion of the CC is commonly reported in linguistic tasks as word generation (Paulesu et al., 2011), but not its posterior section. Notably, the PCC is a central part of the Default Mode Network, and its deactivation occurs when participants are engaged in demanding tasks (Leech et al., 2011; Leech and Sharp, 2014). Thus, the observed groups' difference might be due to higher attentional effort experienced by Deaf-LIS during covert words pronunciation, neuro-anatomically marked by deactivation of the PCC.

Table 2. Group comparison. Location of activation for covert words reading relative to passive viewing of consonants strings, at peaks that were significant at p<0.01 (uncorrected for multiple comparisons).

Approximate location (BAs)	MNI coordinates			
	x	y	z	T value
Deaf-LIS > Hearings				
L middle occipital gyrus (OutBAs)	-25	-65	0	3.8
L middle occipital gyrus (19)	-43	-65	-8	3.44
L inferior occpital gyrus (19)	-36	-65	-8	3.02
L fusiform gyrus (19)	-36	-65	-12	2.81
Deaf-SI > Deaf-LIS				
L posterior cingulate (OutBAs)	-15	-23	36	4.67
L posterior cingulate (23)	-11	-42	28	3.16
L posterior cingulate (23)	-8	-49	19	2.96

Note. Out BAs = Outside defined Bas; Broadman areas have been identified with the Yale MNI to Tal application (Lacadie et al., 2008)

Discussion

The present study was designed to investigate the neuronal basis of covert word reading in a sample of: a) deaf proficient readers who preferentially use Italian Sign Language to communicate, b) deaf proficient readers who preferentially use spoken Italian to communicate and with limited or no knowledge of sign language; c) hearing monolingual controls, proficient readers and with no experience of sign language. The hypothesis was that different neuronal resources would have been recruited by the three groups of participants depending on their auditory experience and preferred mode of communication. Overall, a complex pattern of activation emerged from this study, with relevant between group differences based on their auditory and linguistic experiences.

We first consider the within-group pattern of activations elicited by covert words reading with respect to passive viewing of unpronounceable strings of consonants.

In *hearing participants,* the aforementioned contrast elicited higher activations within superior temporal regions of the right hemisphere. The STG is considered a multisensory integration site, receiving input from unimodal visual and auditory areas, and in the right hemisphere it appears to have a central role in audiovisual and audiomotor speech convergence (Komeilipoor, Cesari & Daffertshofer, 2017), and in dialogical form of inner speech (Alderson-Day et al., 2016). Higher activations have been also reported in middle portion of the right temporal gyrus, previously associated with lexical-semantic processing (Binder, Desai, Graves, & Conant, 2009). This suggests that when covertly reading words, hearing participants automatically access word meaning to facilitate their silent pronunciations.

In *Deaf preferentially using spoken Italian* a rich network of bilateral activations emerged. Activations within the left hemisphere refer to the processing of semantic and lexical information (i.e., medial frontal gyrus, middle temporal gyrus and anterior cingulate), and in the right hemisphere are associated with auditory processing and visual recognition (i.e., right STG and fusiform gyrus) These large pattern of activations might be related to the within group's variability, previously reported also in a behavioral lexical decision task (Barca et al., 2013). In addition, is in line also with the reduced left hemispheric lateralization of the reading-related cortical network reported in deaf individuals (D'Hondt & Leybaert, 2003).

Mode of Acquisition plays an important role in the acquisition of linguistic skills (Wauters, Tellings, van Bon & Mak, 2003), and deaf children with hearing parents (as this group of participants) typically receive insufficient linguistic inputs as parents are not familiar with sign language. Deaf-SI participants had a later exposure to an accessible language (spoken and signed), thus experiencing a period of language deprivation (Rinaldi, Pavani & Caselli, in press). On the contrary, Deaf-LIS participants have been exposed to a natural language (sign language) very early in their life, avoiding the risk of linguistic deprivation. Thus for Deaf-SI, both spoken and sign language might be poorly established, leading to a linguistic processing cost, possibly resulting in the enhanced recruitment of the anterior cingulate cortex (see also MacSweeney et al., 2008). This opens an important field of investigation for future researches.

Both hearing and Deaf-SI showed increased activity in the right STG. We speculate that this might be related to their preference in using a language based on the auditory/verbal modality, based on the finding of increase activation of this region when passively listening to sounds with clear changes in pitch/intonation (Scott et al., 2000). The activation we observe is more posterior than the one reported by Scott et al. (2000). However our limited number of participants and their variability probably makes our cortical localizations less precise than studies with greater sample of participants.

We hypothesized that in *Deaf preferentially using sign language*, covert word reading would result in increased activity in cortical regions involved in hand movements. Our results did not confirm this hypothesis by showing, instead, enhanced activation of supramarginal gyrus and precentral regions of the left hemisphere, typically associated with phonological processing. Anatomically, the SMG has reciprocal connection with ventral premotor cortex and pars opercularis, both involved in articulatory motor planning (Barca et al. 2011; Price, 2012) Enhanced activation of SMG by deaf participants during a rhyming task indicates greater reliance on the articulatory component of speech during phonological processing (MacSweeney, Brammer, Waters & Goswami, 2009). Moreover, together with the angular gyrus, this region is believed to mediate between phonological and orthographic information (Barca et al., 2011; Cornelissen et al., 2009; Pammer et al., 2006; Pammer et al., 2004). In Tunik et al., (2008) activation of the SMG has been implicated in planning goal-oriented hand-object interactions, which does not seem to be directly linked to the production of gestures.

As for the present study, one reason why we did not observe a significant increase in the activation of hand movements areas might be due to a lack of precision in the fMRI temporal information. Kinematic studies underline the dynamic aspects of visual recognitions of objects and words, with online verification processes allowing for correcting response on the fly (Barca, Benedetti & Pezzulo, 2015; Barca & Pezzulo, 2012, 2015; Barca, Pezzulo, Ouellet & Ferrand, 2017). Future studies using techniques with greater temporal resolution such as Magnetoencephalography (Barca et al., 2011; Cornelissen et al., 2009, Urooj et al.,2014) might help to better investigate the dynamics of the cortical network involved in this task.

To explore the effect of hearing status on the neural system supporting covert word reading, we contrasted activation pattern of deaf and hearing participants.

Compared with *Hearing* participants, *Deaf preferentially using sign language* showed enhanced recruitment of lingual and fusiform gyrus which are typically involved in processing verbal material and in language generation tasks (Ghosh, Basu, Kumaran & Khushu, 2010).

Within the fusiform gyrus is place the 'Visual Word Form Area' - VWFA (Dehaene & Cohen, 2011; Martin et al., 2019; McCandliss, Cohen & Dehaene, 2003). The location we have found is more ventral and posterior than its typical MNI location (x = -43, y = -54, z = -12), but some variations between studies have been observed (Barca et al., 2011; Corina et al., 2013). This suggests that Deaf-LIS rely on visual-orthographic information largely than hearings (Price et al., 2003; Price & Devlin, 2011).

The absence of a significant difference when contrasting *Hearing* participants with *Deaf preferentially using spoken Italian* not completely surprising. In our previous work, indeed, these same groups performed similarly in a visual lexical

31

decision task (Barca et al., 2013). However, alternative explanations might account for this similarity. For example, both groups have linguistic experiences mainly in spoken language and these, rather than the knowledge of sign language, appear to modulate the reading network. These needs to be tested in future studies.

Finally, to explore if communication mode (signed versus spoken) modulates the neural system supporting covert word reading, we contrasted the activation pattern of deaf preferentially using sign language and deaf preferentially using spoken language.

Compared to deaf using sign language, those who use spoken language showed higher activation of the cingulate cortex. The dorsal part of the anterior cingulate is anatomically connected with prefrontal and parietal cortex, motor cortex and frontal eye fields, with functions related to error detection, anticipation tasks and attention towards task-relevant stimuli (Weissman et al., 2005). We suggest that the observed groups' difference might be due to higher attentional effort experienced by Deaf-LIS during covert words reading, neuro-anatomically marked by deactivation of the pCC.

Finally, with this same pool of participants, functional connectivity of auditory and linguistic networks is shape by communication mode and linguistic experience (Napolitano et al., 2019). Deaf-LIS have increased activity in intrinsic connectivity within the auditory network (comprising primary and association auditory cortices, associated to action-execution-speech, cognition-language-speech, and perception-audition paradigms) and the fronto-parietal network (corresponding to cognition-language paradigms) with respect to Deaf-SI. Thus, supporting the idea of the existence of a very basic network that acts as a general framework for language processing.

Conclusion

We provided evidence that the cerebral network involved in covert word reading is partially shared by hearings and deaf readers with different auditory experience and communication mode. Yet, we demonstrate that this network does not perform identically across groups, and is modulated by their linguistic experience. Direct comparison between groups have shown enhanced activation of acoustic regions for hearing and deaf using spoken language, and motor regions related to speech-articulation in deaf native signers. Additional differences might be in place when considering dynamics of temporal activation of this network.

Acknowlodegment

The research leading to these results was funded by the European's Community Seventh Framework Programme under grant agreement no. PERG02-GA-2007-224919 to Laura Barca.

***

In our lives there are some turning points, whose value is appreciated only many years after their occurrence.

I met Cristina in the early nineties. She was the speaker of a seminar within my Psychology degree course, and I was among the young students captured by her lively and passionate presentation of the Dual-Route Cascaded model of reading. This is where my journey into Psycholinguistics and Experimental Psychology began.

She played a central role in my scientific training, working side by side in the initial stages, and subsequently promoting my independence with visits to international laboratories. Thanks to her, I also met and collaborated with many authors present in this volume.

I was very fortunate to have followed that seminar and to have Cristina as 'external supervisor' of my graduation thesis, and I join the many colleagues gathered together to celebrate her scientific career.

References

Abutalebi J, GreenDW (2016) Neuroimaging of language control in bilinguals: neural adaptation and reserve. *Bilingualism: Language and Cognition, 19,* 689-698.

Alderson-Day, B. Weis, S. McCarthy-Jones, S., Moseley, P., Smailes, D., Fernyhough, C. (2016). The brain's conversation with itself: neural substrates of dialogic inner speech. *Social Cognitive and Affective Neuroscience,* 110–1201

Aparicio, M., Gounot, D., Demont, E., & Metz-Lutz, M.-N. (2007). Phonological processing in relation to reading: an fMRI study in deaf readers. *NeuroImage, 35*(3), 1303–1316.

Ardila, A., Bernal, B., Rosselli, M. (2016). How localized are language brain areas? A review of Brodmann areas involvement in oral language. *Archives of Clinical Neuropsychology, 31,* 112-122.

Barca L, Bello A, Volterra V, Burani C (2009) Lexical-semantic reading in a shallow orthography: Evidence from a girl with Williams Syndrome. *Reading and Writing. An Interdisciplinary Journal, 23,* 569-588.

Barca, L., Benedetti, F., Pezzulo, G. (2015). The effects of phonological similarity on the semantic categorization of pictorial and lexical stimuli: Evidence from continuous behavioral measures. *Journal of Cognitive Psychology, 28(2),* 159–170

Barca, L., Cornelissen, P., Simpson, M., Urooj, U., Woods, W., & Ellis, A. W. (2011). The neural basis of the right visual field advantage in reading: An MEG analysis using virtual electrodes. *Brain and Language, 118,* 53–71.

Barca, L., Pezzulo, G. (2012). Unfolding visual lexical decision in time. *PloS One, 7(4),* e35932.

Barca, L., Pezzulo, G. (2015). Second thoughts: Continuous and discrete revision processes during visual lexical decision. *PloS One, 10(2),* e0116193

Barca, L., Pezzulo, G. (2017). The influence of communication mode on written language processing and beyond. *Behavioral and Brain Sciences Commentary Invitation, 40.*

Barca, L., Pezzulo, G., Castrataro, M., Rinaldi, P., & Caselli, M. C. (2013). Visual word recognition in deaf readers: lexicality is modulated by communication mode. *PloS One,* 8(3), e59080. doi:10.1371/journal.pone.0059080.

Barca, L., Pezzulo, G., Ouellet, M., Ferrand, L. (2017). Dynamic lexical decisions in French: Evidence for a feedback inconsistency effect. *Acta Psychologica, 180,* 23-32.

Binder, J.R., Desai, R.H., Graves, W.W., Conant, L.L. (2009). Where is the semantic system? A critical review and meta-analysis of 120 functional neuroimaging studies. *Cerebral Cortex, 19,* 2767-2796.

Blamey, P. J. (2003). Development of spoken language by deaf children. In M. Marschark & P. E. Spencer (Eds.), *Oxford handbook of deaf studies, language, and education* (pp. 232–246). New York: Oxford University Press.

Carreiras, M., Mechelli, A., Estévez, A., Price, C.J. (2007) Brain Activation for Lexical Decision and Reading Aloud: Two Sides of the Same Coin? *Journal of Cognitive Neuroscience, 19,* 433-444.

Caselli, M.C., Maragna, S., Volterra, V. (2006). *Linguaggio e Sordità. Gesti, segni e parole nello sviluppo e nell'educazione.* Bologna: Il Mulino.

Corina, D.P., Lawyer, L.A., Hauser, P., Hirshorn, E., (2013). Lexical Processing in Deaf Readers: and fMRI Investigation of Reading Proficiency. *PlosOne, 8,* e54696

Corina, D., Chiu, Y.-S., Knapp, H., Greenwald, R., San Jose-Robertson, L., & Braun, A. (2007). Neural correlates of human action observation in hearing and deaf subjects. *Brain research,* 1152, 111–129. doi:10.1016/j.brainres.2007.03.054.

Cornelissen, P., Kringelbach, M.L., Ellis, A.W., Whitney, C., Holliday, I.E., Hansen, P. (2009). Activation of the Left Inferior Frontal Gyrus in the First 200 ms of Reading: Evidence from Magnetoencephalography (MEG). *PLoS One 4(4):* e5359. doi:10.1371/journal.pone.0005359

Dehaene, S, & Cohen, L. (2011). The unique role of the visual word form area in reading. *Trends in Cognitive Sciences,* 15(6), 254–262. doi:10.1016/j.tics.2011.04.003.

D'Hondt M., Leybaert, J., (2003) Lateralization effects during semantic and rhyme judgement tasks in deaf and hearing subjects. *Brain and Language, 87,* 227–240.

Dietz, N. A., Jones, K. M., Gareau, L., Zeffiro, T. A., & Eden, G. F. (2005). Phonological decoding involves left fusiform gyrus. *Human Brain Mapping, 26,* 81–93.

D'Souza, D., D'Souza, H. (2016). Bilingual Language Control Mechanisms in Anterior Cingulate Cortex and Dorsolateral Prefrontal Cortex: A Developmental Perspective. *Journal of Neuroscience, 36,* 5434-5436.

Gabrieli, J.D.E., Poldrack, R.A., Desmond, J.E. (1998). The role of the left prefrontal cortex in language and memory. *Proc. Natl. Acad. Sci., 95,* 906–913.

Komeilipoor, N., Cesari, P., Daffertshofer, A. (2017).Involvemenet of superior temporal areas in audiovisual and audiomotor speech integration. *Neuroscience, 343,* 276-283.

Lacadie, C.L., Fulbright, R.K., Constable, R.T., Papademetris, X. (2008). More accurate Talairach coordinates for Neuroimaging using nonlinear registration. *Neuroimage, 42(2),* 717-725.

Leech R, Kamourieh S, Beckmann CF, Sharp DJ. (2011). Fractionating the default mode network: distinct contributions of the ventral and dorsal posterior cingulate cortex to cognitive control. *Journal Neuroscience , 31,* 3217–24.

Leech, R., Sharp, D.J. (2014). The role of the posterior cingulate cortex in cognition and disease. *Brain, 137,* 12-32.

MacSweeney, M.,Brammer, M., Waters, D., Goswami, U. (2009). Enhanced activation of the left inferior frontal gyrus in deaf and dyslexic adults during rhyming. *Brain, 132,* 1928-1940.

MacSweeney, M., Waters, D., Brammer, M. J., Woll, B., & Goswami, U. (2008). Phonological processing in deaf signers and the impact of age of first language acquisition. *NeuroImage, 40*(3), 1369–1379.

Marinkovic, K., Dhond, R. P., Dale, A. M., Glessner, M., Carr, V., & Halgren, E. (2003). Spatiotemporal dynamics of modality-specific and supramodal word processing. *Neuron, 38*(3), 487–497.

Martin, L., Durisko, C., Moore, M.W., Coutanche, M.N., Chen, D., Fiez, J.A. (2019). The VWFA is the home of orthographic learning when houses are used as letters. *eNeuro,* 2-13.

Mechelli, A., Friston, K., Price, C.J. (2000). The Effects of Presentation Rate During Word and Pseudoword Reading: A Comparison of PET and fMRI. *Journal of Cognitive Neuroscience, 12, 145-156.*

McCandliss, B.D., Cohen, L., Dehaene, S. (2003). The visual word form area: expertise for reading in the fusiform gyrus. *Trends in Cognitive Neuroscience, 7,* 293-299.

Morford JP, Wilkinson E, Villwock A, Piñar P, Kroll JF (2011) When deaf signers read English: do written words activate their sign translations? *Cognition, 118,* 286–292.

Napolitano, A., Andellini, M., Cannatà, V., Randisi, F., Bernardi, B., Castrataro, M., Pezzulo, G., Rinaldi, P., Caselli, M. C. & Barca, L. (2019, April 15) Analysis of Group ICA functional connectivity of task-driven fMRI: application to language processes in adults with auditory deprivation. *Preprint at https://doi.org/10.31219/osf.io/cnvk8*

Pagliuca G, Arduino LS, Barca L, Burani C (2008) Fully transparent orthography, yet lexical reading aloud: The lexicality effect in Italian. *Language and Cognitive Processes 23*: 422–433.

Pammer, K., Hansen, P. C., Holliday, I., Cornelissen, P. L. (2006). Attentional shifting and the role of the dorsal pathway in visual word recognition. *Neuropsychologia, 44*, 2926-2936

Pammer, K., Hansen, P. C., Kringelbach, M. L., Holliday, I., Barnes, G., Hillebrand, A., Cornelissen, P. L. (2004). Visual word recognition: the first half second. *NeuroImage, 22*(4), 1819–1825.

Pascual-Leone, A., & Hamilton, R. (2001). The metamodal organization of the brain. *Progress in brain research, 134*, 427–445.

Pattamadilok, C., Knierim ,I.N., Kawabata Duncan, K.J., Devlin, J.T. (2010). How does learning to read affect speech perception? *Journal of Neuroscience, 30*, 8435–8444.

Paulesu, E., Goldacre, B., Scifo, P., Cappa, S.F., Gilardi, M.C., Castiglioni, I., Perani, D., Fazio, F. (1997). Functional heterogeneity of left frontal cortex as revealed by fMRI. *Neuroreport, 27*, 2011-7.

Pobric, G., Jefferies, E. & Lambon Ralph, M. A. (2010). Category-specific versus category-general semantic impairment induced by Transcranial Magnetic Stimulation, *Current Biology, 20*, 964-968

Price, C. J. (2012). A review and synthesis of the first 20 years of PET and fMRI studies of heard speech, spoken language and reading. *NeuroImage, 62*(2), 816–847.

Price, C.J., Devlin, J.T. (2011). The interctaive account of ventral occipitotemporal contributions to reading. *Trends in Cognitive Neuroscience, 15*(6), 246-253.

Price, C.J., Winterburn, D., Giraud, A. L., Moore, C. J., & Noppeney, U. (2003). Cortical localisation of the visual and auditory word form areas: A reconsideration of the evidence. *Brain and Language, 86*(2), 272–286.

Reich, L., Szwed, M., Cohen, L., & Amedi, A. (2011). A Ventral Visual Stream Reading Center Independent of Visual Experience. *Current Biology, 21*(5), 363–368.

Rinaldi, P., Caselli, M. C., Onofrio, D., Volterra, V. (2014). Language acquisition by bilingual deaf preschoolers: theoretical, methodological issues and empirical data. In M. Marschark, G. Tang, & H. Knoors (Eds.), *Bilingualism and Bilingual Deaf Education* (pp. 54-73). New York, NY: Oxford University Press.

Rinaldi P., Pavani F., Caselli M.C. (in press). Developmental, cognitive, and neurocognitive perspectives on language development in children who use cochlear implants. *The Oxford Handbook of Deaf Studies in Learning and Cognition.*In M. Marschark, & H. Knoors (Eds.), New York, NY: Oxford University Press

Roswandowitz, C., Kappes, C., Obrig, H., von Kriegstein, K. (2018). Obligatory and facultative brain regions for voice-identity recognition. *Brain, 141*, 234-247.

Seghier, M. L., Neufeld, N. H., Zeidman, P., Leff, A. P., Mechelli, A., Nagendran, A., ... Price, C. J. (2012). Reading without the left ventral occipito-temporal cortex. *Neuropsychologia, 50*(14), 3621–3635.

Shallice T, Kartsounis LD. (1993). Selective impairment of retrieving people's names: a category specific disorder? *Cortex, 29(2)*, 281-91.

Sliwinska, M.W., Khadilkar, M., Campbell-Ratcliffe, J., Quevenco, F., and Devlin J.T. (2012). Early and sustained supramarginal gyrus contributions to phonological processing. *Frontiers in Psychology,* doi: 10.3389/fpsyg.2012.00161

Scott, S., Blanck, C.C., Rosen, S., Wise, R.J.S. (2000). Identification of a pathway for intelligible speech in the left temporal lobe. *Brain, 123*, 2400-2406.

Tunik, E.,Lo, O., Adamovich, S. (2008). Transcranial Magnetic Stimulation to the Frontal Operculum and Supramarginal Gyrus Disrupts Planning of Outcome-Based Hand–Object Interactions. *The Journal of Neuroscience, 28(53):*14422–14427

Urooj, U., Cornelissen, P., Simpson, M., Wheat, K.L., Woods, W., Barca, L., Ellis, A.W. (2014). Interactions between visual and semantic processing during object recognition revealed by modulatory effects of age of acquisition. *Neuroimage, 87*, 252-264

Wauters, L. N., Tellings, A. E., Van Bon, W. H., & Mak, W. M. (2007). Mode of acquisition as a factor in deaf children's reading comprehension. *Journal of Deaf Studies and Deaf Education, 13*(2), 175-192.

Weissman, D. H., Gopalakrishnan, A., Hazlett, C. J., & Woldorff, M. G. (2005). Dorsal Anterior Cingulate Cortex Resolves Conflict from Distracting Stimuli by Boosting Attention toward Relevant Events. *Cerebral Cortex, 15(2)*, 229–237.

Wheat, K. L., Cornelissen, P. L., Frost, S. J., & Hansen, P. C. (2010). During visual word recognition, phonology is accessed within 100 ms and may be mediated by a speech production code: evidence from magnetoencephalography. *The Journal of neuroscience: the official journal of the Society for Neuroscience, 30*(15), 5229–5233.

Appendix.

1.1 Participants
Deaf participants filled in an Anamnestic Questionnaire, providing self-report information about years of education, experience with Italian Sign Language (LIS), frequency and context of LIS use, and family characteristics (e.g., deaf relatives).
Deaf with a preference for Italian Sign language (Deaf-LIS). The group consisted of 7 deaf signers with severe to profound bilateral sensorineural hearing loss (71+ dB in the better ear). They had learned LIS in a family context (deaf children from deaf parents) or at school (deaf children from hearing parents), in a 'naturalistic' fashion, and within 3 years of age. They primarily use LIS for communication and adopt it in different social contexts (at home, at school, with friends). They also frequently use spoken language, mainly accompanied by corresponding signs. They attended mainstream schools with a 'communication assistant' who used LIS to communicate and to convey school subjects.
Deaf with a preference for Spoken Italian (Deaf-SI). The group consisted of 7 deaf participants with severe to profound bilateral sensorineural hearing loss (71+ dB in the better ear). All of them were born from hearing parents and primarily used spoken language to communicate. They made limited or no use of LIS. Those who know LIS had learned it after 15 years of age. They all use hearing aids regularly and none had a cochlear implant when data were collected. They had attended mainstream schools with teachers who used spoken Italian to communicate and to convey school subjects.
Both groups underwent speech therapy during their school years.

1.2 Apparatus, image acquisition, processing and analysis
All measurements were performed on a clinical Philips 1.5T Nova Dual Achieva scanner (Philips Medical Systems, The Netherlands) using the body coil transmission and an 8-element phased-array head coil as receiving coil. Neural activity was measured by acquiring $T2^*$-weighted images with BOLD contrast. Functional images were acquired in three runs, each of 120 volumes comprising 30 axial slices of 5 mm thickness and an in-plane resolution of 4 mm. The scanner protocol was as follows: TR=2000 ms, TE=30 ms, flip angle=90°, FOV=256 mm×256 mm, matrix=64×64. The protocol comprised a T1-weighted 3DMPRAGE sequence (TR = 2400 ms, TE = 3.13 ms, flip angle = 8°, voxel size = 1.0 ×1.0 × 1.0 mm).
Stimuli were presented by using STIM2, NeuroScan 2003 software version. In the scanner, visual stimuli were delivered by an MR-compatible set of goggles.
FMRI SPM5 (http://www.fil.ion.ucl.ac.uk/spm/) was used for the analysis of the fMRI data. Functional scans were motion corrected, resliced, normalized at 2x2x2

mm3 to a standard brain (Montreal Neurological Institute space) via the high resolution structural image, and smoothed with an 8 mm FWHM Gaussian kernel to allow for group analysis.

A general linear model (GLM) was used to retrieve functional information and analyze individual results contrasting task. Parameters for the three spatial directions and for the three rotations determined by motion corrections have been considered as repressors and added to the GLM. The resulting contrast files (words > nonwords) were tested against zero with a one-sample t-test and tested among the groups with a paired T-test applying false discovery rate correction (Benjamini & Hochberg, 1995). A p-value of 0.01 was considered the significance threshold.

The BioImage suite Web have been used to confirm anatomical locations of MNI coordinates (Lacadie et al., 2008)

Visual word recognition is sensitive to social information extracted from vocal cues

Simone Sulpizio[1]*, Fabio Fasoli[2]

[1]Faculty of Psychology, Vita Salute San Raffaele University, Italy.

[2]University of Surrey, United Kingdom

*sulpizio.simone@hsr.it

Abstract.

We investigated here whether information about speaker's social identity implicitly inferred from voice affects the recognition process of stereotypical words. We focused on sexual orientation, a social category that has not yet received attention in the context of word and message processing. In one experiment, we asked participants to listen to either a gay- or heterosexual-sounding voice reading a brief text; then, participants completed a lexical decision task including stereotypically-gay, -heterosexual, or -neutral words, half with a positive and half with a negative valence. Results showed that the exposure to gay (vs. heterosexual) sounding speaker differently affected stereotypes activation in male and female listeners: Male listeners were slower in recognizing negative stereotypically-gay traits, but only after being exposed to a gay-sounding speaker; instead, female participants were faster in recognizing negative traits when listened to the heterosexual-sounding speaker, and faster in recognizing stereotypically-heterosexual traits when listening to the gay-sounding speaker. The findings are discussed at the light of stereotyping, social communication theories, and psycholinguistic theories of word processing.

In this paper, we present the results of a study that combines something extremely close to Cristina's work – that is visual word recognition studied by means of lexical decision task – with something that is pretty far from her research – that is speaker's social identity conveyed by vocal cues. Thus, this study combines where the first author of the present work comes from with where he is (eventually) going to; this trip would not have started without meeting Cristina. We hope she will enjoy – although she may be a bit horrified! – in reading this paper.

The study of visual word recognition goes back to the middle of last century and has been always interested in unveiling the psychological mechanisms at the basis of this unique culturally-acquired ability. Along the years, most has been understood on how the recognition system works – e.g., that the system is flexible enough to occur regardless of how the words are presented handle with different forms (e.g., difference in fonts, shapes, colors, e.g., Dehaene, Jobert, Naccache, Ciuciu, Poline, Le Bihan, & Cohen 2004), that it is sensitive to orthographic properties of the language (e.g., Traficante & Burani, 2014; Velan & Frost, 2007), or that it automatically accesses to orthographic, phonological, and semantic information during word processing (e.g., Amenta, Marelli, & Sulpizio, 2017; Barca, Bello, Volterra, & Burani, 2009; Harm & Seidenberg, 2004). Thanks to this prolific research activity, current theorists can precisely describe the mechanisms that allow readers to go from the printed information to their mental linguistic representations (for a recent vision, see Grainger, Dufau, & Ziegler, 2016).

Despite this precision in describing the mechanisms underlying visual word recognition, research on this topic shares with the rest of the psycholinguistic research an important limitation: it has developed in the mainstream tradition of cold cognition (see, e.g., Jacobs, 2015). This tradition comes from the modular approach to cognition, which assumes that the language system, as any other cognitive system, is an encapsulated module and can thus be investigated apart from many aspects of everyday communicative situations. However, as social animals, we use language as our main tool to communicate our thoughts and share our feelings. Therefore, understanding how the language system interacts with social information is one of the challenges of contemporary psycholinguistic research (for a similar view, see van Berkum, 2018). The present work aims to face this challenge by investigating whether and to what extent social cues may affect the recognition of a printed word. In particular, we will examine how social information coming from vocal cues affects recognition of stereotypical words associated to the speaker's identity. In so doing, we considered a social category, namely sexual orientation, that has not yet received attention in the context of word and message processing. Before jumping into the details of our research, we will briefly review the literature on auditory gaydar and stereotyping.

Sexual Orientation, Gaydar and Stereotyping

Stereotyping occurs automatically (Devine, 1989) as shown by numerous implicit (RTs based) measures where participants' strength of associations between social groups and stereotypes have been tested (see Fazio & Olson, 2003; Kawakami, Young, & Dovidio, 2002). Such stereotypes are activated as soon as a person is categorized as a member of a given group (Fiske & Neuberg, 1990). However, there are social groups that are somehow 'ambiguous'. One of these 'ambiguous' social categories is sexual orientation (Tskhay & Rule, 2013). We cannot be sure about someone's sexual orientation until the person self-discloses as gay, lesbian, bisexual or of any other sexual orientation. Indeed, people tend to assume the majority of people to be heterosexual unless cues of a gay sexual orientation, usually gender atypicality, occur (Lick & Johnson, 2016). Research on categorization of sexual orientation – a process also known as *gaydar* – has shown that people use different cues to guess if someone is gay or heterosexual (Rule, 2017). Accuracy in gaydar judgments varies depending on the cue under investigation. Sexual orientation judgments from facial cues have been found to be quite accurate, and better for female than male targets (Rule, 2017). Recent studies considering voices have instead shown that accuracy is generally low (see Kachel et al., 2017; 2018) and that listeners are somehow hesitant in judging speakers as gay (Sulpizio et al., 2015). Nevertheless, auditory gaydar judgments do not seem to be random. Across cross-linguistic studies, Sulpizio and colleagues have shown poor accuracy in judging sexual orientation of male (Sulpizio et al., 2015) and female speakers (Sulpizio et al., 2019), but also that listeners were consistent in judging some speakers as gay/lesbian-sounding, regardless of their actual sexual orientation. These judgments were linked to specific acoustic features, as well as to common stereotypes about the 'gay voice'. For instance, formant frequencies and length of some vowel, nasality, and spectral features of sibilant /s/ have been found to be linked to perceived gay sexual orientation (Sulpizio et al., 2015; Kachel et al., 2018). At the same time, if a male speaker sounds feminine and a female speaker sounds masculine, they are more likely to be categorized as gay and lesbian, respectively (Munson, 2007). Hence, all in all this literature has shown that voice is taken as a cue of sexual orientation although it does not always lead to accurate gaydar judgments.

Perceiving someone as gay/lesbian, or simply perceiving them as gay/lesbian-sounding, can affect listeners' first impressions. As a matter of fact, male speakers who sound gay are perceived as having personality traits and personal interests (e.g., hobbies and study subjects) that are typically assessed as feminine (Fasoli, Maass, Paladino, & Sulpizio, 2017). This also extends to inferences about the type of job they do and their health status: listeners imagine gay-sounding speakers as better suited for typically feminine professions (see Fasoli & Maass, 2018) and as more likely to suffer from diseases that are typically associated with women (e.g., anorexia; Fasoli, Maass, & Sulpizio, 2018) than heterosexual-sounding speakers. People seems to appreciate the consistency between stereotype and vocal cues as a process to maintain shared knowledge. As a matter of fact, people prefer actors whose voice matches with their stereotypic characteristics (e.g., a feminine character with a gay/sounding voice (Fasoli, Mazzurega, & Sulpizio, 2017) and show stronger stigmatization towards individuals who sound gay but disclose as heterosexual (Gowen & Britt, 2006).

So far, literature has provided evidence that a gay-sounding speaker is stereotyped as feminine and having feminine interests. However, it remains to understand whether sexual orientation conveyed by vocal cues activated stereotypes generally. In this study we tested whether the mere exposure of gay- vs. heterosexual-sounding speakers made gay stereotypes more accessible. At this regard, a study by Rule, Macrae, and Ambady (2009) is particularly relevant here. In this study, the authors found that the exposure to 10 gay rather than 10 heterosexual male faces facilitated the activation of stereotype consistent words in a lexical decision task. In particular, participants exposed to gay faces were faster in recognizing concepts stereotypically associated with gay men (e.g., rainbow). However, whether vocal cues of sexual orientation have a similar impact has not been examined yet. Hence, we tested the prediction that listening to gay-sounding speakers would activate gay stereotypes. In particular, we expected that participants would be faster in recognizing gay stereotypes as words after being exposed to the voice of a gay- than a heterosexual-sounding speaker. Moreover, we expanded previous research by testing the effect of voice on activation of both positive and negative stereotypes. A limit of Rule et al.'s (2009) study was that the valence of target words was not considered. It is important to examine the interplay between evaluative (valence) and conceptual (stereotype) associations that may emerge from the activation of specific social categories (see Wittenbrink, Judd, & Park, 2001). Research has shown that exposure to stimuli referring to social minority groups activate negative evaluation and stereotyping automatically (see, e.g., Dovidio, Kawakami, Johnson, Johnson, & Howard, 1997; Fazio, Jackson, Dunton, & Williams, 1995). Gay men are target of social prejudice in Italy (Eurobarometer, 2015) and therefore, along with stereotyping, negative evaluations are likely to arise.

Method

Participants.
Thirty-nine students (18 males, mean age: 22.89, sd: 3.01) from the University of Trento took part in the experiment. All participants were Italian native speakers, with normal or correct-to-normal vision. After being informed about the study, participants provided written informed consent to their participation. The study was approved by the ethical committee of the University of Trento.

Materials.
Stimuli for lexical decision. Thirty Italian words were selected as experimental stimuli; the words were taken from Fasoli et al.'s (2017) Experiment 1. The words were 10 stereotypically-gay traits (*romantico, curato, sensibile, premuroso, creativo, insicuro, effeminato, emotivo, pettegolo, malizioso*), 10 stereotypically-heterosexual traits (*vigoroso, forte, deciso, pratico, dominante, aggressivo, violento, rozzo, prepotente, arrogante*), and 10 stereotypically-neutral traits (*onesto, fiducioso, saggio, fidato, simpatico, avaro, tirchio, formale, imbroglione, noioso*) . In each category, 5 were positive and 5 were negative traits. Stereotypically-gay, -heterosexual, and -neutral traits were matched on written frequency (extracted from SUBTLEX-IT database, Crepaldi, Keuleers, Mandera, & Brysbaert, 2013) and length (all $ps > .1$). Thirty legal non-word fillers were created and matched with words on length.

Speakers and audio materials. Two male speakers, pronouncing the same short text (see appendix), were employed. Speakers were young adult Italian native speakers (age range: 26/33), recruited in the northeast regions of Italy. Speakers were chosen from a database of voice samples provided by Sulpizio et al. (2015) and on the basis of ratings of Experiment 2 where perceived sexual orientation was assessed on a scale from 1 (exclusively heterosexual) to 6 (exclusively homosexual). The heterosexual speaker was overall perceived as heterosexual-sounding ($M = 1.69, SD = .85$) and the gay speaker was perceived as gay-sounding ($M = 5.24, SD = 1.39$). Hence, we used stimuli that referred to a gay and a heterosexual speaker whose voice conveyed their sexual orientation.

Speakers were uttering a short text (~40 seconds) that referred to the structure of DESY, a research center in Austria. The text was taken from an online italo-german magazine titled 'Contrasto'. The text was 204 words long. We selected this text as its content was neutral and unrelated to sexual orientation stereotypes.

Design & Procedure.
The experiment had a 2 (Speaker Sexual Orientation: gay vs. heterosexual) x 3 (Type of traits: stereotypically-gay vs. -heterosexual vs. -neutral) x 2 (Trait valence: positive vs. negative) x 2 (Participants gender: female vs. male). Speaker' sexual orientation and participants' gender were between-participants factors, whereas

44

type of traits and trait valence were within-participants factors.

The experiment started with the listening session: Participants seat in front of the screen and were asked to wear headphones and to listen to a brief audio track; no information was given either on the speaker or on the message content. Each participant listened to either the gay or the heterosexual speaker. At the end of the listening phase, participants removed the headphones and were instructed to perform the lexical decision task. Participants were informed that strings of letters would appear on the screen and their task consisted in indicating whether the string referred to a meaningful word or a non-existing word as quickly and accurately as possible. Target stimuli consisted of different type of traits and non-words as illustrated above. Stimuli were presented in a random order in one block. Each trial of this task started with a fixation cross, presented for 300 ms in the center of the screen; the fixation was followed by a short blank screen of 200 ms; then a stimulus appeared in the same position and was presented until the participant responded or for a maximum of 1500ms; finally, there was an inter-stimulus interval of 800 ms. Responses were given by pressing either the "m" or "x" keys on the keyboard; key selection was counterbalanced across participants. The experiment started with a brief practice session and was run using E-Prime software (Psychology Software Tools, Pittsburgh, PA; www.pstnet.com).

Results

Errors were few (2.56% of data points) and were not analyzed. Reaction times (RTs) of correct responses were analyzed using mixed-effects models (Baayen, Davidson, & Bates, 2008); the models were fitted using *lmer* function (*lmerTest* package) in R software. Participants and items were entered as random factors, whereas speaker's sexual orientation (gay vs. heterosexual), type of traits (stereotypically-gay vs. -heterosexual vs. -neutral), trait valence (positive vs. negative), and participants gender (female vs. male) were included as fixed factors. Results are reported in Figure 1.

The model showed a main effect of trait valence ($F = 5.47$, $p = .02$), and a two-way interaction between trait valence and speaker sexual orientation ($F = 7.93$, $p = .004$). More interestingly, the four-way interaction was also significant ($F = 6.06$, $p = .002$).

To further inspect the interaction, data were split for participants gender; for female participants the new analysis showed a main effect of trait valence ($F = 4.27$, $p = .04$), and two-way interactions, one between speaker sexual orientation and trait valence ($F = 6.59$, $p = .01$), and one between speaker sexual orientation and type of traits ($F = 3.38$, $p = .03$).

45

Figure 1. Mean RTs for female (upper panel) and male participants (lower panel) for the different experimental conditions. Vertical bars indicate standard errors. H = heterosexual; N = neutral; G = gay.

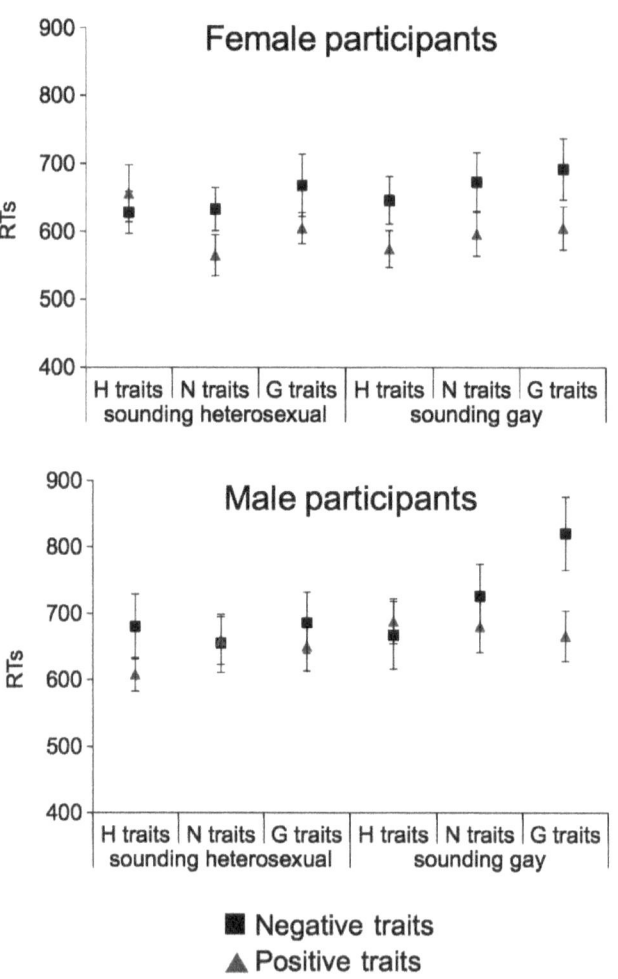

The first two-way interaction suggested that female participants were faster in recognizing negative traits when the speaker was heterosexual (M = 643 SD = 162) than gay (M = 670, SD = 181) while similar RTs emerged for positive traits ($M_{heterosexual}$ = 608, SD = 149 vs. M_{gay} = 592, SD = 133). The second interaction suggested that female participants were faster in reacting to stereotypically-heterosexual traits when the speaker was gay (M = 610, SD = 142) than heterosexual (M = 642, SD = 162), whereas they were faster in reacting to stereotypically-neutral traits when the speaker was heterosexual (M = 599, SD = 142) than gay (M = 635, SD = 171); more similar RTs emerged for stereotypically-gay traits spoken by gay (M = 648, SD = 175) and heterosexual speakers (M = 636, SD = 162).

For male participants, the analysis showed a main effect of trait valence (F = 5.91, p = .02), and a three-way interaction between trait valence, type of traits, and speaker sexual orientation (F = 4.18, p = .01); further inspection of this interaction showed that, when categorizing stereotypically-gay traits, the exposure to a gay-sounding speaker slowed down the recognition of negative (beta = 128.18, st. err. = 62.19, t = 2.06, p = .05), but not positive gay stereotypical traits (t < 1, p > .7); speaker's sexual orientation had no effect on the categorization of stereotypically-heterosexual or -neutral traits (all ps > .1).

Discussion

Results showed that voice influenced word recognition and stereotype activation. However, the way in which exposure to gay (vs. heterosexual) sounding speaker affected stereotypes activation was different for male and female listeners. Male speakers were slower in recognizing negative stereotypically-gay traits. This result suggests that, instead of a generalized activation of stereotypical traits, male participants somehow selectively inhibit stereotypical negative gay traits. It is possible that the exposure to a gay-sounding speaker may have increased awareness of our participants' heterosexual identity. Since manhood is particularly important for heterosexual men (Vandello et al., 2008), their identity may become salient when primed with cues of gay sexual orientation (see Fasoli et al., 2018). As a consequence, this may have made negative gay stereotypes less readily.

Female participants showed an unexpected pattern of results. They were faster in recognizing negative traits when the speaker was heterosexual. This might be related to an ingroup differentiation strategy (Tajfel & Turner, 1979): heterosexual female participants would be more prompted to activate negative concepts after being exposed to an outgroup member, namely a heterosexual man. At the same time, female participants were faster in recognizing stereotypical traits associated to heterosexual men when listening to gay speakers. Being primed with cues of gay sexual orientation may activate those traits that reflect the social norms, namely being heterosexual (see Kimmel, 1997).

Overall, these findings suggest that heterosexual men and women are influenced by gay-sounding voices in different ways. The motives that lead to activation of different types of stereotypical traits are distinct and are likely to be related to listeners' own identity and intergroup dynamics (Bosson & Vandello, 2013). Nevertheless, encountering someone who sounds gay seems to affect what type of semantic information is available. It seems to be easy to stereotype someone who sounds gay (Fasoli et al., 2017; 2018), but general stereotype activation seems to be more complex. In this study, participants were not asked to judge the speaker but to recognize words in a simple lexical decision task. What our results seems to suggest is that this task and the availability of stereotypical traits is very much related to who the listeners are (e.g., their gender). Our results are different from those by Rule et al. (2009) who did not find any gender difference, but simply showed a stereotype activation when individuals were primed with cues of sexual orientation. The reason of such a difference may be due to the type of cues under investigation, with vocal cues being processed differently from facial cues. Moreover, listening to someone may activate a context where participants are engaged and 'prepared' for a social interaction. Research has shown that different communication schemas allow individuals to accommodate or diverge to the person they are interacting with (see Communication Accommodation Theory, Giles, 2016); by highlighting their similarity and differences, participants create a context for an exchange.

In a broader perspective, the present findings challenge the classic view of a cold language processing and suggest that the early mechanisms of visual word recognition are sensitive to and take into account task-irrelevant social information while processing the written words. Thus, the impression that listeners get about the speaker from her/his voice contributes to the recognition process by either facilitating or interfering with the categorization of the written stimulus as word or not.

It may be tentatively suggested that, similarly to what has been reported for face (e.g., Freeman & Ambady, 2011; Zhang, Morris, Cheng, & Yap, 2013), voice activates information about social categories that is assumed to be in contact with lexico-semantic knowledge by means of direct links. In other words, hearing a voice immediately triggers the categorization of speaker's identity; this information spreads into the lexico-semantic system, affecting the availability of identity-related lexico-semantic knowledge. This hypothesis is based on an adaptation of the dual-route model of spoken word recognition proposed by Sumner and colleagues (2013). The model assumes that speech signal is simultaneously mapped into linguistic and social representation; since these two types of information are connected via bidirectional links, social representation is able to modulate word processing at pre-lexical and semantic levels.

To conclude, the present study has shown that social information about the speaker's identity extracted from voice is automatically used by listeners to perform a simple lexical decision task. This finding suggests that social content may exert a direct influence on language processing and advocate for a research that aims to investigate how social processing and language processing interact with each other.

**

Addendum. I met Cristina by chance, in a sunny Capitoline day; now, I think I was extremely lucky to meet her. She has been my first mentor and thanks to her I could move my first steps into psycholinguistics. However, what I probably appreciate more about her is her scientific and personal integrity, as well as her terrific ability to discreetly give good personal advises and support. Thanks Cristina, I have really enjoyed working with you!

Simone Sulpizio

References

Amenta, S., Marelli, M., & Sulpizio, S. (2017). From sound to meaning: Phonological-to-Semantcs mapping in visual word recognition. *Psychonomic Bulletin & Review, 24,* 887-893.

Barca, L., Bello, A., Volterra, V., & Burani, C. (2010). Lexical-semantic reading in a shallow orthography: Evidence from a girl with Williams syndrome. *Reading and Writing: An Interdisciplinary Journal, 23,* 569-588.

Baayen, R.H., Davidson, D.J., & Bates, D.M. (2008). Mixed-effects modeling with crossed random effects for subjects and items. *Journal of Memory and Language, 59,* 390-412.

Vandello, J.A., & Bosson, J.K. (2013). Hard won and easily lost: A review and synthesis of theory and research on precarious manhood. *Psychology of Men & Masculinity,* 14(2), 101-113.

Crepaldi, D., Keuleers, E., Mandera, P., & Brysbaert, M. (2013). SUBTLEX-IT. Retrieved from http://crr.ugent.be/subtlex-it/

Dehaene, S., Jobert, A., Naccache, L., Ciuciu, P., Poline, J.B., Le Bihan, D., & Cohen, L. (2004). Letter binding and invariant recognition of masked words: behavioral and neuroimaging evidence. *Psychological Science, 15,* 307-313.

Dovidio, J.F., Kawakami, K., Johnson, C., Johnson, B., & Howard, A. (1997). On the nature of prejudice: Automatic and controlled processes. *Journal of Experimental Social Psychology, 33,* 510-540.

Eurobarometer, S. (2015). 83. Spring 2015. *Public opinion in the European Union. Regime to access: http://ec. europa. eu/public_opinion/archives/eb/eb83/eb83_first_en. pdJ.*

Fasoli, F., Maass, A., Paladino, M.P., & Sulpizio, S. (2017). Gay- and Lesbian-Sounding Auditory Cues Elicit Stereotyping and Discrimination. *Archives of Sexual Behavior, 46,* 1261-1277.

Fasoli, F., Maass, A., & Sulpizio, S. (2018). Stereotypical disease inferences from gay/lesbian vs. heterosexual voice. *Journal of Homosexuality, 65,* 990-1014.

Fasoli, F., Mazzurega, M., & Sulpizio, S. (2017). When characters impact on dubbing: the role of sexual stereotypes on voice actor/actress' preferences. *Media Psychology, 20,* 450-476.

Fazio, R.H., Jackson, J.R., Dunton, B.C., & Williams, C.J. (1995). Attitudes and social cognition. *Journal of Personality and Social Psychology, 69,* 1013-1027.

Fazio, R.H., & Olson, M.A. (2003). Implicit measures in social cognition research: Their meaning and use. *Annual Review of Psychology, 54,* 297-327.

Fiske, S.T., & Neuberg, S.L. (1990). A continuum of impression formation, from category-based to individuating processes: Influences of information and motivation on attention and interpretation. In *Advances in Experimental social Psychology* (Vol. 23, pp. 1-74). Academic Press.

Freeman, J.B., and Ambady, N. (2011). A dynamic interactive theory of personal construal. *Psycholological Review, 118,* 247-279.

Gowen, C.W., & Britt, T.W. (2006). The interactive effects of homosexual speech and sexual orientation on the stigmatization of men: Evidence for expectancy violation theory. *Journal of Language and Social Psychology, 25,* 437-456.

Grainger, J., Dufau, S., & Ziegler, J.C. (2016). A vision of reading. *Trends in Cognitive Sciences, 20,* 171-179.

Harm, M.W., & Seidenberg, M.S. (2004). Computing the meanings of words in reading: Cooperative division of labor between visual and phonological processes. *Psychological Review, 111,* 662-720.

Jacobs, A.M. (2015). Towards a neurocognitive poetics model of literary reading. *Cognitive Neuroscience of Natural Language use,* 135-59.

Kachel, S., Simpson, A.P., & Steffens, M.C. (2017). Acoustic correlates of sexual orientation and gender-role self-concept in women's speech. *The Journal of the Acoustical Society of America, 141,* 4793-4809.

Kachel, S., Simpson, A.P., & Steffens, M.C. (2018). "Do I Sound Straight?": Acoustic Correlates of Actual and Perceived Sexual Orientation and Masculinity/Femininity in Men's Speech. *Journal of Speech, Language, and Hearing Research, 61,* 1560-1578.

Kawakami, K., Young, H., & Dovidio, J.F. (2002). Automatic stereotyping: Category, trait, and behavioral activations. *Personality and Social Psychology Bulletin, 28,* 3-15.

Kimmel, M. S. (1997). Masculinity as homophobia: Fear, shame and silence in the construction of gender identity. In M. M. Gergen & S. N. Davis (Eds.), *Toward a new psychology of gender* (pp. 223–242). New York: Routledge.

Lick, D.J., & Johnson, K.L. (2016). Straight until proven gay: A systematic bias toward straight categorizations in sexual orientation judgments. *Journal of Personality and Social psychology, 110,* 801-817.

Munson, B. (2007). The acoustic correlates of perceived masculinity, perceived femininity, and perceived sexual orientation. *Language and Speech, 50,* 125-142.

Rule, N. O. (2017). Perceptions of sexual orientation from minimal cues. *Archives of Sexual Behavior, 46,* 129-139.

Rule, N.O., Macrae, C.N., & Ambady, N. (2009). Ambiguous group membership is extracted automatically from faces. *Psychological Science, 20,* 441-443.

Sulpizio, S., Fasoli, F., Maass, A., Paladino, M.P., Vespignani, F., Eyssel, F., Bentler, D. (2015). The sound of voice: Voice-based categorization of speakers' sexual orientation within and across languages. *PLoS ONE, 10:* e0128882.

Sulpizio, S., Fasoli, F., Antonio, R., Eyssel, F., Paladino, M.P., & Diehl, C. (2019). Auditory gaydar: Perception of sexual orientation based on female voice. *Language and Speech.* Advanced online publication.

Sumner, M., Kim, S.K., & King, K.B. M. (2013). The socially weighted encoding of spoken words: a dual-route approach to speech perception. *Frontiers in Psychology, 4.*

Tajfel, H., Turner, J.C., Austin, W.G., & Worchel, S. (1979). An integrative theory of intergroup conflict. *Organizational identity: A reader,* 56-65.

Traficante, D., & Burani, C. (2014). List context effects in languages with opaque and transparent orthographies: a challenge for models of reading. *Frontiers in Psychology, 5,* 1023.

Tskhay, K. O., & Rule, N.O. (2013). Accuracy in categorizing perceptually ambiguous groups: A review and meta-analysis. *Personality and Social Psychology Review, 17,* 72-86.

Velan, H., & Frost, R. (2007). Cambridge University versus Hebrew University: The impact of letter transposition on reading English and Hebrew. *Psychonomic Bulletin & Review, 14,* 913-918.

Van Berkum, J.J.A. (2018). *Language comprehension, emotion, and sociality: Aren't we missing something? In Rueschemeyer, S. A. & Gaskell, G. (Ed.). Oxford Handbook of Psycholinguistics. (pp. 644-669). Oxford: Oxford University Press.*

Vandello, J.A., Bosson, J.K., Cohen, D., Burnaford, R.M., & Weaver, J.R. (2008). Precarious manhood. *Journal of Personality and Social Psychology, 95,* 1325-1339.

Wittenbrink, B., Judd, C.M., & Park, B. (2001). Evaluative versus conceptual judgments in automatic stereotyping and prejudice. *Journal of Experimental Social Psychology, 37, 244-252.*

Zhang, S., Morris, M.W., Cheng, C.Y., & Yap, A.J. (2013). Heritage-culture images disrupt immigrants' second-language processing through triggering first-language interference. *Proceedings of the National Academy of Sciences, 110,* 11272-11277.

Morpheme-based spelling in Italian: evidence from children with dyslexia and normally developing readers

Paola Angelelli[1]*, Chiara V. Marinelli[1],

[1] Department of History, Society and Human Studies, Lab. of Applied Psychology and Intervention, University of Salento, Lecce, Italy

*paola.angelelli@unisalento.it

Abstract

In the present paper we reviewed studies on spelling acquisition in transparent orthographies, such as Italian, in order to highlight spelling strategies beyond the phoneme-to-grapheme (P-G) conversion and whole-word orthographic procedure. In fact, also in transparent orthographies, where an efficient P-G mapping may be sufficient to reach a good level of accuracy, sublexical units larger than phonemes/graphemes, such morpheme or cluster of letters, but below the whole-word level, play an important role in learning to spelling. Even if children are explicitly taught the application of P-G correspondences, implicitly they develop a sensitivity to sublexical larger sound-to-print relationships. Moreover lexical and sublexical processes interact very early supporting spelling acquisition in both children with typical and atypical development.

Spelling is the main activity in primary school: it is estimated that 30 to 60% of a child's school-day is focused on writing activities. Moreover developmental spelling deficits, together with dyslexia, are the most common learning disturbances, affecting 5-15% of the population (APA, 2013).

The present review analyze the interaction between lexical and sublexical factors influencing spelling acquisition in children learning a transparent orthography, as well as the implicit learning of intra- and inter-word regularities that might facilitate spelling. Most of Italian studies were developed thanks to the sapient guide of Cristina Burani, who solicited our critical view on the traditional dual-route models of spelling and shared with us her profound knowledge of psycholinguistic research across languages.

The dual-route model assumes the existence of at least two processes for spelling (Patterson, 1986): A lexical process, which relies on accessing word-specific memory and may be semantically mediated or may involve direct connections between phonology and orthography, and a sublexical process, based on P-G mapping. Notarnicola, Angelelli, Judica, & Zoccolotti (2012) analyzed the spelling performance of Italian 1st- to 8th- grade normally developing children and found that both lexical and sublexical procedures are available since the first year of schooling, but showed different developmental trends: accuracy on pseudoword spelling (i.e., sublexical spelling) showed a rapid increase followed a plateau around third-fourth grade, while accuracy on ambiguous words (i.e., lexical processing) was generally lower and increased also successively. These data indicate an earlier and more rapid development for the sublexical procedure and a more gradual acquisition in the case of the lexical procedure. Coherently, cross-linguistic studies highlighted that the more regular the writing system the more children rely on sublexical processing (for an English-Italian comparison see Marinelli, Romani, Burani, & Zoccolotti, 2015) and this procedure is acquired rapidly and efficiently, with ceiling effect for pseudoword spelling by the 1st or 2nd grade (for a review Caravolas, 2004). Conversely, lexical processing in Italian was less efficient than in English, because English is more consistent when larger units are processed (Marinelli et al., 2015).

However, according to more recent models (Perry & Ziegler, 2004), individuals do not only spell by applying P-G mapping or retrieving whole words, but are also sensitive to the frequency of relationships between sound and spelling at various grain sizes below the whole word. With the sapient guide of Cristina Burani we have examined if also Italian children with both typical and atypical development use, not only P-G mapping and lexical processing, but also larger grain size (also if smaller than the whole words).

Morphological effect

Morphemes constitute intermediate units between the single grapheme and the whole word. In opaque orthographies, morphological information favours spelling accuracy, expecially when the mapping is inconsistent (for a review see Pacton and Deacon, 2008): for example in English morphological conventions assist the transcription of regular past tens verbs ending in -*ed* although its pronunciation might be /d/, /t/ and /Id/.

However, a facilitatory effect of morphology in spelling was detected also in transparent orthographies. Lehtonen and Bryant (2005) examined Finnish, a richly inflected language with highly transparent orthography. Authors used words in which target clusters of letters (the sequences LL and SS) might be either root (unbound morpheme) or inflection. Note that in Finnish, case inflections are a more prominent part of morphology than derivation, because they occur in nouns, adjectives, pronouns and numerals. Results highlighted that by the end of the 1st grade children began to spell letter clusters better in case inflections than in word roots, which suggested early sensitivity to the morphological structure of words in spelling. Similar results were found for pseudowords: letter clusters occurring in endings corresponding to case inflections were spelled with greater accuracy than those occurring in pseudo-roots, suggesting that case like endings prompted morphological parsing during spelling. Authors suggest that the facilitation arises because the children's mental lexicon is organized in morphemes and case inflections are solidly acquired and represented in their lexicon due to the high frequency with which they occur. This in turn helps the subsequent parsing of words into their constituent morphemes, favouring the spelling.

With regard to Italian, we conducted a first study (Angelelli, Marinelli, & Burani, 2014) in which 3rd and 5th grade normally developing children read aloud and spelled words and pseudowords varying for the presence of morphemic constituents. We found that Italian children benefitted from the presence of morphological constituents in spelling newly encountered stimuli: in fact facilitation was limited to pseudowords made up of existing morphemes (that were read and spelled more accurately than non-morphemic pseudowords), irrespective of school level (with a somewhat higher advantage for younger than for older children). This result is a genuine morphological effect, rather than a generic "word-likeness" effect, because we carefully controlled for familiarity of the chunks constituting the pseudowords (in the non-morphological set, final orthographic sequences were as frequent as suffixes in the root + suffix pseudoword set). Furthermore, morphological and non-morphological pseudowords were matched for bigram frequency. Then, in developing readers, exposure to these frequently occurring sound-to-print units with a meaning, allows morphemes to become relatively independent spelling units.

The morphological facilitation in spelling was confirmed also in Italian children with dyslexia and dysgraphia (Angelelli, Marinelli, De Salvatore & Burani, 2017): they took advantage as controls in spelling pseudowords with morphological constituents as well as low frequency words made up of high frequency morphemes (while words with low-frequency morphemes were spelled comparably to non-derived words). This finding is coherent with a recent Spanish study (Suàrez-Coalla, Martínez-García, Cuetos, 2017), showing that children with dyslexia benefitted from a high frequency base to read and spell unfamiliar stimuli.

In sum, also in transparent orthographies typically developing children as well as dyslexic/dysgraphic children benefit of the presence of morphemes when spelling stimuli without a lexical entry (such as pseudowords or low frequency words), highlighting that they exploit orthographic regularities which reflect the statistical distribution in their orthography, and that they might use large-sized processing units in spelling.

Neighbourhood effect

Further evidence for lexical influence on the sublexical spelling comes from studies that exploited the effect of phonological word neighbours on pseudoword spelling. A phonological neighbour is commonly defined as a stimulus (word or pseudoword) that sounds similar to a target word because it shares with the target all the phonemes in the same positions except one (e.g., Landauer, & Streeter 1973). In a first study examining the effect of neighbourhood size in spelling, Tainturier and colleagues (2013) studied French-speaking proficient adults. Authors dictated pseudowords which had a phonological word neighbour spelled with a low-probability P-G (uncommon) mapping (e.g., /o/ transcribed as AUD). The authors found that low-probability P-G mappings were used significantly more often in spelling pseudowords that had a phonological lexical neighbour containing that spelling (e.g., participants spelled /o/ as AUD when they heard /krepo/, a pseudoword which has a word neighbour, CRAPAUD, /krapo/) than in spelling pseudowords with no neighbours. The use of uncommon mappings in pseudoword spelling is an evidence of lexical influence of the neighbour: upon auditory presentation of a pseudoword, phonological word neighbours are activated in the phonological lexicon (for a review Gow, 2012) and this phonological activation triggers the spelling of the word neighbours, which in turn affects the spelling of the target pseudoword. Interestingly, the authors found that the magnitude of lexical influence increased with the lexical frequency of the neighbours.

Evidence supporting the influence of lexical orthographic process on pseudoword spelling is present also in studies on typically developing children (e.g., Bosse, Valdois & Tainturier, 2003; Martinet, Valdois & Fayol, 2004). For example, Bosse and colleagues (2003) dictated pseudowords to French 1st to 5th graders and found that children spelled pseudowords by analogy to known words. An analogy effect was found even in first graders, when children knew how to spell the reference word. Similarly, Martinet and colleagues (2004), by strictly controlling for the lexical database to which children had been exposed, found an analogy effect in pseudoword spelling after only three months of reading instruction.

In our study, we explore the effect of orthographic neighbourhood on pseudoword spelling of Italian 3rd and 5th grade typically developing children (Angelelli, Notarnicola, Marcolini, & Burani, 2014). Stimuli were short and long pseudowords derived from high- or low-frequency words by changing either the first or the fourth letter (early vs late diverging pseudoword). Results highlighted that long pseudowords were spelled as accurately as short stimuli if derived from high-frequency words. Furthermore, early diverging pseudowords, if derived from high-frequency words, were spelled more accurately than other types of stimuli.

Overall, with an other experimental paradigm, we find again that, similarly to what occur in opaque orthographies, lexical information is used to support sublexical spelling also by young learners of a transparent orthography such as Italian.

Lexical priming effect and distributional knowledge

Children might implicit learn intra- and inter-word regularities, along with explicit learning of P-G correspondences. In order to investigate if children use this probabilistic cues for spelling and reading and if this knowledge interacts with whole-word lexical processing, Italian children with dyslexia and controls were examined longitudinally from 3rd to 5th grade (Marinelli, Cellini, Zoccolotti & Angelelli, 2017). To this purpose, we capitalized on cases of inconsistent G-P mappings that present a clear asymmetry in the occurrence of each orthographic option. For example the phonemic group [kw] might be transcoded as CU or QU (e.g., CUOCO /'kwɔko/, (cook) vs. QUOTA /'kwɔta/, (share)), while the sound [tʃe]; [ʃe] might be written as CIE, SCIE or CE, SCE (e.g., CIECO /'tʃɛko/, (blind) vs. CECO / 'tʃɛko/, (Czech), SCIENZA /'ʃɛntsa/, (science) vs ADOLESCENZA /adoleʃ'ʃɛntsa/, (adolescence)). Both cases present an asymmetry in the occurrence of each orthographic option: the phonemic groups [tʃe], [ʃe], [kw] are more frequently spelled as CE, SCE, QU (they appear in 77.2 to 82.8% of cases according to the databases of Marconi, Ott, Pesenti, Ratti & Tavella, 1993, and De Mauro, 1989, respectively) as compared to CIE, SCIE, CU, which were more atypical (from 22.8% to 17.2% of cases). Results highlighted that dyslexic children, as well as control children, had acquired the distributional knowledge and preferred typical transcription to atypical one. In particular, dyslexic children used this probabilistic information for compensating their difficulty when lexical processing was not available, as in the case of low frequency words. This allow to ensuring the correct spelling at least of words with typical mappings, even in the absence of lexical representation.

Other studies addressed this issue by examining lexical priming effects on pseudoword spelling (e.g., Folk & Rapp, 2004): the spelling of a target pseudowords[2] in a related priming condition (a relate words previously heard word) and in a control condition (a free-spelling task of pseudowords or an unrelated priming condition). Results in English speaking adults showed that pseudoword spelling could be affected by the orthography of the previously heard word. Lexical priming effects on pseudoword spelling were also found in Italian skilled adults (Barry & De Bastiani, 1997): participants were prone to spell pseudowords with the orthographic segment that occurred in the prime word they had previously heard. Therefore, also in a consistent orthography such as Italian, sublexical spelling was influenced by whole-word lexical processing, at least in skilled adults.

With Cristina Burani we carried out a series of experiments on lexical priming in Italian typically developing children and children with dyslexia. In experiment 1, a dictation task of words including inconsistently spelled phonemes with typical or atypical spelling. In experiment 2, we examined lexical priming effect on pseudoword spelling. Words of experiment 1 were used as prime words for the spelling of pseudowords containing the same inconsistently spelled segments (related priming condition). As a control, we examined the effect of unrelated words (with consistent G-P mapping) on pseudoword spelling (unrelated priming condition). In both experiments, we manipulated not only the type of transcription of the ambiguous phonemes (typical vs atypical), but also the frequency of the prime word. In a first study (Angelelli et al., 2017), we studied Italian normally developing children attending 1st, 2nd and 4th grade; in the second study (Marinelli et al., in preparation) we extended the experimental paradigm to dyslexic children. In the first study, both experiments showed that sensitivity to the distributional properties of orthography is an early acquisition in Italian. Already in 1st grade, typical P-G mappings were spelled more correctly than atypical ones in spelling both words and pseudowords. In spelling atypical segments, higher accuracy were obtained in higher grades, probably when whole-word lexical representations start to be established. Moreover, lexical priming effect were reported. In related prime condition, children spelled pseudowords with the same transcription present in prime words: the atypical transcription was used more frequently after atypical prime words respect to control condition; similarly related primes with typical

---

[2] In these experiments the target spelling of a pseudoword referred to the particular spelling of the pseudoword's vowel that corresponded to the spelling of the same vowel in the prime word.

mappings maximized the use of typical mapping in pseudoword spelling, respect to the unrelated priming condition. This result highlight that whole-word lexical processing interacts with knowledge of sublexical regularities and influence pseudoword spelling. Note that the lexical priming effects is larger for older children with respect to 1st graders, due to the higher lexical processing in the former, and for high frequency primes, because stronger lexical activation occurs respect to low frequency prime words.

Similarly to control children, also dyslexic children preferred typical transcriptions (respect to atypical one) in pseudoword spelling. Moreover, coherently with Marinelli et al. (2017), they showed difficulty in spelling low-frequency words with unpredictable mapping only if containing an atypical transcription (while words with typical transcriptions were spelled comparably to controls). However, dyslexic children showed smaller lexical priming effect respect to controls, probably for the limitations of their orthographic lexicon (Angelelli, Marinelli & Zoccolotti, 2010). Then, with a different experimental paradigm, we confirmed that also children with dyslexia are sensitive to distributional properties of the orthography and sublexical knowledge supports lexical spelling processing.

In conclusion we hope to have contributed to a better understanding of the different processes underlying spelling in a transparent orthography and we thank again Cristina Burani for the time, dedication and interest toward this issue.

References

American Psychiatric Association (2013). Diagnostic and Statistical Manual on Mental Disorders. 5th ed. (DSM-5). Washington, DC: American Psychiatric Press.

Angelelli, P., Marinelli, C.V., & Burani, C. (2014). The effect of morphology on spelling and reading accuracy: a study on Italian children. *Frontiers in Psychology*, 5, 1-10.

Angelelli, P., Marinelli, C.V., De Salvatore, M., & Burani, C. (2017). Morpheme-based reading and spelling in Italian children with developmental dyslexia and dysgraphia. *Dyslexia: An International Journal of Research and Practice*, 23, 387-405.

Angelelli, P., Marinelli, C.V., Putzolu, A., Notarnicola, A., Iaia, M., & Burani, C. (2017). Learning to spell in a language with transparent orthography: distributional properties of orthography and whole-word lexical processing. *Quarterly Journal of Experimental Psychology*, 71, 704-716.

Angelelli P., Marinelli C.V., Zoccolotti P. (2010). Single or dual representation for reading and spelling? A study on Italian dyslexic and dysgraphic children. *Cognitive Neuropsychology*, 27, 305- 333.

Angelelli, P., Notarnicola, A., Marcolini S., & Burani, C. (2014). Interaction between the lexical and sublexical spelling procedures: A study on Italian primary school children. In. Special issue Ed. by M.A. Pinto, & S. D'Amico: Lexical access: studies on monolingual and plurilingual subjects at different developmental stages, *Journal of Applied Psycholinguistics*, XIV, 2. ISSN 1592-1328 (Print); ISSN 1724-0646 (Electronic)

Barry, C., & De Bastiani, P. (1997), Lexical priming of nonword spelling in the regular orthography of Italian. *Reading and Writing: An Interdisciplinary Journal*, 9, 499-517.

Bosse, M.L., Valdois, S., & Tainturier, M.J. (2003). Analogy without priming in early spelling development. *Reading and Writing: An Interdisciplinary Journal*, 16, 693-716.

Caravolas, M. (2004). Spelling development in alphabetic writing systems: A cross-linguistic perspective. *European Psychologist*, 9, 3–14.

De Mauro, T. (1989). VELI. Vocabolario elettronico della lingua Italiana [VELI. Electronic dictionary of Italian language]. Milano: IBM Italia.

Folk, J. R., & Rapp, B. (2004). Interaction of lexical and sublexical information in spelling: Evidence from nonword spelling. *Applied Psycholinguistics*, 25, 565–585.

Gow, D.W. Jr. (2012). The cortical organization of lexical knowledge: a dual lexicon model of spoken language processing. *Brain and Language*, 121, 273-288.

Landauer, T.K. & Streeter, L. A. (1973). Structural differences between common and rare words: failure of equivalence assumptions from theories of word recognition.*Journal of Verbal Learning and Verbal Behaviour*, 12, 119-131.

Lehtonen, A., & Bryant, P. (2005). Active players or just passive bystanders? The role of morphemes in spelling development in a transparent orthography. *Applied Psycholinguistics,* 26, 137–155.

Marconi, L., Ott, M., Pesenti, E., Ratti, D., & Tavella, M. (1993). Lessico elementare: Dati statistici sull'italiano scritto e letto dai bambini delle elementari [Elementary lexicon: Statistical data for Italian written and read by elementary school children]. Bologna: Zanichelli.

Marinelli C.V., Cellini P., Zoccolotti P., & Angelelli P. (2017). Lexical processing and distributional knowledge in sound-spelling mapping in a consistent orthography: A longitudinal study of reading and spelling in dyslexic and and typically developing children. *Cognitive Neuropsychology*, 34, 163-186.

Marinelli, C.V., Romani, C., Burani, C., & Zoccolotti, P. (2015). Spelling acquisition in English and Italian: A cross-linguistic study. *Frontiers in Psychology - Educational Psychology*, 6:1843.

Martinet, C., Valdois, S., & Fayol, M. (2004). Lexcial orthographic knowledge develops from the beginning of literacy acquisition. *Cognition*, B11-B22.

Notarnicola, A., Angelelli, P., Judica, A. & Zoccolotti, P. (2012). The Development of spelling skills in a shallow orthography: The case of the Italian language. *Reading and Writing: An Interdisciplinary Journal*, 25, 1171-1194.

Pacton, S., & Deacon, S. H. (2008). The timing and mechanisms of children's use of morphological information in spelling: a review of evidence from English and French. *Cognitive Development,* 23, 339–359.

Patterson, K. E. (1986). Lexical but nonsemantic spelling. *Cognitive Neuropsychology*, 3, 341-367.

Perry, C., & Ziegler, J. C. (2004). Beyond the two-strategy model of skilled spelling: Effects of consistency, grain size, and orthographic redundancy. The Quarterly Journal of Experimental Psychology, 57A, 325-356.

Suàrez-Coalla, P., Martínez-García, C., & Cuetos, F. (2017). Morpheme-Based Reading and Writing in Spanish Children with Dyslexia. *Frontiers in Psychology*, 8:1952

Tainturier, MJ, Bosse, ML, Roberts D.J., Valdois, S., & Rapp, B. (2013). Lexical neighborhood effects in pseudoword spelling. *Frontiers in Psychology*, 4, 862.

Psycholinguistic factors and global components in developmental dyslexia

Pierluigi Zoccolotti[1,2*†], Chiara V. Marinelli[3], Marialuisa Martelli[1], Donatella Spinelli[4]
Maria De Luca[5]

[1] Department of Psychology, Sapienza University of Rome, Rome, Italy.

[2] Institute of Cognitive Sciences and Technologies, ISTC-CNR, Rome, Italy.

[3] Department of History, Society and Human Studies, Lab. of Applied Psychology and Intervention, University of Salento, Lecce, Italy.

[4] Department of Movement, Human and Health Sciences, University of Rome "Foro Italico", Rome, Italy.

[5] Neuropsychology Unit, IRCCS Fondazione Santa Lucia, Rome, Italy.

*pierluigi.zoccolotti@uniroma1.it

†Acknowledgement: We have written this chapter as a tribute to Cristina Burani to thank her for her advice and insight, which has inspired our research on developmental dyslexia over the past decade.

Abstract

A large literature (spanning from reading to perceptual, cognitive, attentional and motor processes) has tried to pinpoint the factors causing developmental dyslexia. As a means to identify the core deficit in dyslexia we referred to two models of individual differences in speeded tasks: The Rate and Amount Model (RAM; Faust, Balota, Spieler, & Ferraro, 1999) and the Difference Engine Model (DEM; Myerson, Hale, Zheng, Jenkins, & Widaman 2003). By controlling for the over-additivity effect (i.e., the fact that in raw data differences between typically developed readers and dyslexics tend to grow as a function of the task difficulty over and above the effect of specific condition manipulations), it was possible to establish that much of the widespread group differences in reading tasks could be accounted for by a single global factor. Evidence indicated that this global deficit concerned the processing of orthographic stimuli (independent of their lexical quality) but not the naming of pictures or the analysis of stimuli in the auditory modality. Accordingly, we propose that the global factor responsible for dyslexics' slowing concerns a deficit in forming a pre-lexical representation of a visually presented orthographic string. By controlling for the influence of this global factor we could also examine the specific role of psycholinguistic factors in ways not confounded by the over-additivity effect.

Evidence indicated that children with dyslexia were selectively sensitive to the length of the words. Lexical processing was present in children with dyslexia as indicated by the presence of lexicality and frequency effects, sensitivity to morphological structure and facilitation of large N-size. Still, there was also evidence that children with dyslexia were inefficient in activating low-frequency entries in the lexicon, indicating that lexical organization in these children is similar to that of skilled readers but may be underdeveloped. Further research is needed to clarify whether the presence of an underdeveloped lexicon should be viewed as a late occurring consequence of the pre-lexical deficit in orthographic processing.

Open questions in the study of developmental dyslexia

In spite of over a century of research, interpreting developmental dyslexia poses a number of important challenges. Current theories propose a variety of possible causes spanning from deficits in phonological representations (e.g., Swan & Goswami, 1997) to deficiencies in magnocellular processing (e.g., Stein, 2001), as well as deficits in automatization possibly linked to impaired cerebellar processing (Nicolson, Fawcett, & Dean, 2001). Indeed, this large spread of hypotheses derive from evidence indicating that, compared to children without a reading deficit, children with dyslexia show impaired performance on a large variety of tasks on quite different areas of processing.

For example, it has been reported that various perceptual tasks distinguish between children with and without a reading problem (although failures to replicate some of these effects have been reported, a point which will not be fully developed here). A short list includes: problems in eye movement programming (Bucci, Brémond-Gignac, & Kapoula, 2008), increased visual crowding (Martelli, Di Filippo, Spinelli, & Zoccolotti, 2009), decreased visual span (Bosse, Tainturier, & Valdois, 2007), temporal processing (Farmer & Klein, 1995), and deficits in movement perception (Raymond & Sorensen, 1998). Several studies also indicated problems in visuo-spatial attention (e.g., Vidyasagar & Pammer, 2010; a deficit actually reported to be multimodal; e.g., Facoetti et al., 2010) or executive function (Varvara & Varuzza, 2014). Furthermore, several cognitive tasks reliably distinguish between children with and without a reading defect, including rapid automatized naming (e.g., Denckla & Rudel, 1976) as well as several phonological and meta-phonological tasks (for a review see Melby-Lervåg, Lyster, & Hulme, 2012), including phonemic awareness (Bruck, 1992), sensitivity to rhyme (Bryant, MacLean, & Bradley, 1990) and short-term memory (Swanson, Zheng, & Jerman, 2009). While some of the early reports on basic acoustic processing deficits (e.g., Tallal, 1980) have not been confirmed (for a review see Hämäläinen, Salminen, & Leppänen, 2013), more recent evidence indicates significant deficits in amplitude modulation (rise-time; Goswami, 2011). Furthermore, there is also a long tradition of research examining deficits in speech perception (e.g., Lieberman, Meskill, Chatillon, & Schupack, 1985) even though evidence is mixed as to the conditions under which a deficit reliably emerges (for a discussion see Ramus & Ahissar, 2012). Notably, deficits are not constrained to perceptual and cognitive dimensions and there are several reports indicating the presence of difficulties in fine motor tasks (e.g., Fawcett & Nicolson, 1995) and even deficits in posture maintenance (e.g., Pozzo, Vernet, Creuzot-Garcher, Robichon, Bron, & Quercia, 2006).

Similar difficulties are encountered when examining which reading conditions best signal the core deficit in a given child. Usually, children with dyslexia are impaired across various target stimuli, including reading texts as well as lists of high- and low-frequency, regular and irregular words as well as non-words. The one possible exception to this generalization is the identification of single letters, since several reports indicate sparing of letter processing in children with dyslexia (e.g., Katz & Wicklund, 1972), even in conditions that control for the overall difficulty of the tasks involved (Martelli et al., 2009).

Thus, a large literature has tried to identify which conditions mark most clearly the reading deficit but, again, this determination has proven difficult. One line of research has underscored the centrality of a deficit in reading non-words (Rack, Snowling, & Olson, 1992; van Ijzendoorn & Bus, 1994; Hermann, Matyas, & Pratt, 2006); however, as spelled out more clearly below, serious criticisms have been put forward to this proposal. Alternatively, it has been claimed that different children may be characterized by selective difficulties with different stimulus materials, such as a deficit in reading non-words versus a deficit in reading irregular words (Castles & Coltheart, 1993). Still, the latter disturbance, referred to as surface dyslexia, has been the object of much controversy. This is centered on the question as to whether it represents a selective deficit or merely points to a delay in reading acquisition (e.g., Stanovich, Siegel, & Gottardo, 1997).

This brief overview clearly underscores the difficulty of formulating a single, unitary model of developmental dyslexia considering the variety of problems reported both in terms of the phenomenology of the disturbance and of its causal origin. Thus, there seems to be the need for a method which allows to place the large variety of evidenced deficits in a reasonable context. For example, Goswami (2015) has proposed that many sensory deficits (such as impaired visual attention span or visuospatial attention) are presumably due to the lower level of reading exercise, which impact on child cognitive functions. Indeed, children with dyslexia do not like to engage themselves in reading activities and, as an effect, other things being equal, tend to read less (Stanovich & Cunningham, 1992). In this view, there would be one or few nuclear deficit(s) at the origin of the reading deficit but, by the time children are tested, the reduced exposure to print would produce a variety of difficulties, such as those briefly outlined above. In this line of thought, poor performance in reading irregular words may be seen as a consequence rather than a cause of dyslexia (e.g., Stanovich et al., 1997) although direct supporting evidence on this is lacking (e.g., Peterson, Pennington,& Olson, 2013). As put forward by Goswami (2015), this may also apply to poor performance in other tasks, even less strictly linked to reading experience, such as perceptual span or spatial attention.

The reading level-match approach: rationale and criticisms

Goswami (2015) proposes a number of approaches which may help in disentangling nuclear from associated deficits in dyslexia. However, by and large, the method used most often in this perspective is the reading level-match design. In this paradigm, children with dyslexia are compared in some specific tasks not to typically developing children of the same chronological age but to children with the same level of reading skills (and thus of a younger age). The general idea in this perspective is that, if children with dyslexia are not impaired with regard to skilled children with the same reading age (but typically younger), the deficit is presumably only due to a "delay" in acquisition (e.g., due to reduced exposure to print). By contrast, if children with dyslexia are also impaired in comparison to children with the same reading age, their difficulty in the specific task (such as reading non-words) can be interpreted as a selective "deviance". This latter finding would then be reasonably taken into account in developing a model of dyslexia. Overall, the reading level-match design well expresses the need to have a method for distinguishing, among the several deficits associated with developmental dyslexia, those which are critical for interpreting the disorder.

There is an extensive literature using the reading level-match design. However, there is also a large literature discussing the potential pitfalls of this paradigm (for a discussion on the methodological assumption of this method see for example Mervis & Klein-Tasman, 2004). Here, we focus on a critical methodological issue which makes the use of this paradigm prone to important biases. In particular, it has been demonstrated that the reading level-match design rests on the (unlikely) assumption that the critical variables object of investigation show a homogeneous development (van den Broeck & Geudens, 2012). However, if that is not the case, this method will significantly bias the outcome of the comparison; in particular, this is the case for the well-known finding that children with dyslexia are impaired in reading non-words when compared to children matched for reading level (van den Broeck & Geudens, 2012).

This finding is usually interpreted to indicate that the crucial deficit in reading is due to a phonological deficit since non-words are typically read with reference to the non-lexical phonological routine and no access to the lexicon (Hermann et al., 2006; Rack et al., 1992; van Ijzendoorn & Bus, 1994;). However, van den Broeck and Geudens (2012) showed that this is an artifact. In particular, this is due to the fact that, among typically developing children, performance on non-words develops more slowly and generates more inter-individual variability than performance in reading words. Each of these two characteristics would be sufficient by itself in generating an apparent selective deficit in reading non-words if the comparison is based on a control group matched for reading level. So, the deficit entirely disappears if one uses z scores based on chronological age norms or more complex designs that do not rest on the described critical assumptions (such as the "*state trace*" design; van den Broeck and Geudens, 2012).

An approach based on models of individual differences in speeded tasks

Overall, the lesson from the literature on the reading level-match design is that the method is ineffective in separating the core deficit from non-specific deficits in dyslexia. In developing a research program on developmental dyslexia, we looked for an alternative approach to quantify the reading deficit.

One interesting source is given by models that try to account for individual differences in performance in speeded tasks, as measured by reaction times (RTs). There is a large literature on the use of RTs (often vocal RTs) in measuring a variety of performances and, in fact, it is well known that these are governed by well-established laws: thus, (1) RT distributions are skewed to the right and (2) this skew increases with task difficulty and (3) the distribution spread increases with the mean (Wagenmakers & Brown, 2007). Thus, these relationships need to be incorporated in any model of RT performance.

One interesting model aimed to account for individual differences in RT performance ("*rate and amount model*" or RAM) was developed by Faust et al. (1999). According to these authors, the performance on a given task is largely due to the interaction between two components. One, referred to as rate, is associated with the general "cognitive speed" of the individual; another, named amount, refers to the level of processing required to perform a given task (or difficulty). These two factors interact multiplicatively to produce the performance of an individual on a given task. Overall, the joint effect of the rate and amount factors indicate the role of a global factor which accounts for the performance of individuals across a large variety of cognitive speeded tasks. However, these authors also consider the possibility that additional factors, specific to a given experimental manipulation, contribute to the performance (and provide some statistical methods to detect this contribution). Thus, it is possible that an individual (or a group) is specifically impaired on a given task, even after the individual rate of processing and the general difficulty of a task are taken into account (but note that task specificity cannot be established with reference to analyses on raw data only). Thus, this approach has the potential to identify selective deficits after controlling for the global factor which influences performance. Selective deficits on a specific task may emerge over and above the effects of a global factor.

Another model which focusses on the individual differences in RT performance was developed by Myerson et al. (2003) and is referred to as "*difference engine model*" or DEM. While the RAM model focusses on separating the global and specific sources of variability in individual performance, the DEM model specifically aims to define the characteristics of the global components in the data. In particular, it proposes that individual performance can be ascribed to two sources of variability: a sensory-motor compartment and a cognitive-decisional compartment. The first adds a constant to the individual performance (the time necessary to sensory processing and motor program) and may well be (almost) spared even in groups who show reduced rate of processing, as in the case of older adults or individuals with different pathologies, such as Alzheimer or Parkinson disease (Myerson et al., 2003). The second one indicates the portion of the response time which corresponds to the central cognitive processing. Typically, this compartment generates multiplicative differences among critical groups as a function of task difficulty. Myerson et al. (2003) developed methods to separate the sensory-motor and cognitive compartments and made explicit predictions on the characteristics of the relationships between condition means and variability (i.e., standard deviations, SDs) across groups varying for different general levels of processing.

A clear example is provided by the effect of aging. It is well known that older individuals tend to be slower than younger adults across a large variety of tasks (e.g., Cerella, 1985; Verhaeghen & Cerella, 2002). Myerson et al. (2003) review evidence indicating that the difference between the groups refers only to the cognitive-decisional compartment while that in the sensory compartment is minimal (but for an alternative perspective see Ratcliff, Thapar, Gomez, & McKoon, 2004). Critically, older individuals are slower and more variable but the relationship (slope) between the average group performance and the inter-individual variability is the same. Thus, the SDs (variability) of the mean performances (condition means) of younger and older individuals in various tasks can be fitted in a graph by a single line. This indicates that the basic relationship (or law) that governs the global components of the responses is independent of group as well as cognitive domain (a point which will be developed below).

Overall, the two models may provide interesting and complementary information in analyzing the individual differences in performance in RT tasks.

Studying reading speed in Italian children with dyslexia

We set out to study the characteristics of developmental dyslexia in Italian, a language characterized by a very regular orthography. It is well known that for a long while the literature has seen the prevalence of studies on English (referred to as "anglocentrism" by Share, 2008). English is a very irregular orthography and this may account for the great interest in understanding the reading difficulties in children learning such a difficult orthography. The typical method in examining English-speaking children is that of comparing the accuracy of a group of children with dyslexia in reading lists of critical stimuli with that of a matched group of typically developing children. To this aim, it is critical that this latter group of children shows a sufficient level of reading errors to allow for a group comparison. However, in very regular orthographies, such as Italian or Finnish, the accuracy level of the typically developing children rapidly reaches an asymptote such that by third or fourth grade, reading errors are extremely rare or absent. Thus, apart from the more general problems of adapting models developed in English to more regular orthographies, there is a very basic methodological difficulty in examining reading in languages such as Italian, German or Finnish because accuracy measures are based on a very low number of errors. The alternative option is that of examining measures of reading speed, as originally advocated by Wimmer (1993).

Thus, in planning a research program on the characteristics of reading difficulties in Italian children with dyslexia, we chose to adopt an approach based on RT measures. Note that the use of RTs is actually what is typically adopted in psycholinguistic studies on adults (independent of language). In pursuing this project, we asked for the collaboration of Cristina Burani who joined our research efforts by providing her expertise to develop the critical questions on the nature of developmental dyslexia. The studies described below owe much of their development to Cristina's competence, insight and dedication.

Overall, using RT measures in our studies on the characteristics of developmental dyslexia in Italian had two major advantages. On the one hand, we could base our studies on children on the large literature on psycholinguistic studies on reading in adults; this was indeed our original focus of research. On the other hand, becoming increasingly aware of the potential of RAM and DEM models applied to dyslexia, we set out to interpret individual performance in reading tasks based on these models. Thus, our general aim was to uncover which factors underlie the deficit shown by children with dyslexia. Critically, at least in principle, our research approach allowed partitioning the possible causes without adjudicating *a priori* whether the reading deficit should be ascribed to a selective, specific source or, by contrast, it should be interpreted in terms of a global, non-task specific impairment.

Influence of psycholinguistic factors on the reading performance of Italian children with dyslexia: a line of research

Early in our research project we tried to examine which word factors modulated most clearly the performance of children with dyslexia.

First, we systematically tested the clinical impression that children with dyslexia have a selective difficulty with longer orthographic stimuli. Evidence along these lines was also available in terms of the eye movement pattern during reading (De Luca, Borrelli, Judica, Spinelli, & Zoccolotti, 2002; De Luca, Di Pace, Judica, Spinelli, & Zoccolotti, 1999;). Thus, children with dyslexia showed saccades of a similar amplitude in reading both short and long words; as an effect of this, the number of fixations was linearly related to the number of letters in the word (as well as non-word). By contrast, typically developing readers adjusted their visual inspection as a function of the stimulus characteristics; thus, with longer words, they made saccades of a larger amplitude while, with shorter words, they made smaller saccades (De Luca et al., 2002).

In various RT studies, we examined the effect of stimulus length on the word reading of children with dyslexia (De Luca, Barca, Burani, & Zoccolotti, 2008; Spinelli, De Luca, Di Filippo, Mancini, Martelli & Zoccolotti, 2005; Zoccolotti, De Luca, Gasperini, Judica, & Spinelli, 2005;). The results indicated a strong dependence of RTs upon stimulus length in children with dyslexia at all ages tested (i.e., third, sixth and seventh grade; De Luca et al., 2008; Spinelli et al., 2005;); by contrast, the responses of typically developing readers were heavily dependent upon this factor in first grade (Zoccolotti et al., 2005), but the influence progressively reduced so that, by sixth grade, only a marginal effect of length was detectable (De Luca et al., 2008).

The strong influence of length indicates a tendency for children with dyslexia to read by a sequential analysis of the target stimulus. We note here that this tendency has been reported independent of orthographic regularity. Thus, in a study comparing German- and English-speaking children, Ziegler, Perry, Ma-Wyatt, Ladner, and Schulte-Körne (2003) noted that the RTs in word reading of children with dyslexia were similarly sensitive to word length in both languages. The sensitivity to word length may indicate a tendency to process by partitioning the stimulus through the non-lexical route. This would be consistent with the working hypothesis that children would suffer from a deficit in lexical processing (or surface dyslexia). Note that this hypothesis (which we proposed in our original study of these children; Zoccolotti, De Luca, Di Pace, Judica, Orlandi, & Spinelli, 1999) would be inconsistent with the prevailing view that the key deficit in dyslexia concerns the non-lexical, phonological processing. However, it should be added that other interpretations of the effect of length are also possible; in particular, the enhanced sensitivity to stimulus length might indicate an impairment in early parallel orthographic analysis (e.g., Barton, Hanif, Eklinder Björnström, & Hills, 2014; van den Boer, de Jong, & Haentjens-van Meeteren, 2013;).

To detect the role of lexical versus sub-lexical processing in the reading of children with dyslexia, we carried out a study in which two markers of these processing routines were used, i.e. the frequency effect and the grapheme contextuality effect.

It is well known that word frequency is the critical factor affecting lexical organization (for a review of evidence on Italian see Barca, Burani, & Arduino 2002). As for grapheme contextuality, Italian has a number of graphemes whose pronunciation requires the processing of the subsequent vowel (hereto-after, "complex" graphemes). Thus, "c" is pronounced as /k/ when followed by "a", "o" or "u" (such as in "casa", transl.: house) while it is pronounced as /tʃ/ when followed by "e", or "i" (such as in "cima", transl.: top). By contrast, for a number of graphemes (hereto-after, "simple" graphemes), such as "p", the pronunciation is univocal (such as "palo", transl.: pole). It has been proposed that the locus for this effect is in the non-lexical reading routine. Consistently, it has been reported that, all other factors being equal, words that contain complex graphemes (such as c and g) are read more slowly (and less accurately) than words without complex graphemes, but only in the case of low-frequency words (Burani, Barca, & Ellis, 2006). This is interpreted in terms of the need of sequential processing characteristic of the non-lexical routine in the processing of these stimuli (Burani et al., 2006). By contrast, no effect is present in the case of high frequency words, which is interpreted as due to the holistic processing of the lexical routine. The grapheme contextuality effect was found in adults (Burani et al., 2006) as well as in typically developing readers (Barca, Ellis, & Burani, 2007). Thus, it appeared that examining the influence of the grapheme contextuality effect could provide critical information on the activity of the non-lexical phonological route.

The results of the study indicated that children with dyslexia showed a large frequency effect (both in RTs and in terms of reading accuracy; Barca et al., 2006). Furthermore, in keeping with the performance of typically developing children, children with dyslexia showed the expected pattern for the grapheme contextuality effect: thus, words containing complex graphemes were read more slowly (and less accurately) than words containing only simple graphemes but, as expected, this difference was detected only in the case of low-frequency words.

Thus, this study did not provide a clear indication as to the specific involvement of the lexical or non-lexical routine in modulating the dyslexic deficit of Italian children with dyslexia. In fact, the presence of a rather large frequency effect indicated a spared lexical organization (though acquiring further evidence on this issue seemed certainly in order); furthermore, the presence a grapheme contextuality effect limited to low-frequency words indicated the expected efficiency of the non-lexical routine. So, the question remained as to "*where are the differences*" between children with and without a reading deficit, as spelled out in the title of our paper (Barca et al., 2006).

Overall, these early studies indicated that children with dyslexia were particularly sensitive to word length but also showed the lexical effect of word frequency and the non-lexical effect of the grapheme contextuality effect expected in typically developing readers. How could we frame this pattern of results?

One interesting observation is that children with dyslexia were generally much slower than controls across all experimental conditions. Thus, in studies of word length a quite large RT difference was actually present even for the shortest words presented (e.g., 3—letter words in the case of Spinelli et al. 2005; 4—letter words in the case of De Luca et al., 2008). As indicated above, the presence of a seemingly general speed deficit may significantly influence the performance across different conditions as well as affect the actual size of the group effects found. Faust et al. (1999) refer to this as the "over-additivity" effect; this indicates that, when studying slow and fast groups of individuals, any difference between two critical experimental conditions will be numerically greater in the slower group over and above the specific influence of the experimental manipulations.

This was indeed the picture which emerged in the results at that time. Thus, for example, the difference between high- and low-frequency words was actually larger in children with dyslexia than in matched control children. This indicated the possibility that at least part of the differences between these two groups of children would be due to a general or "global" factor contributing to the performance of children by producing multiplicative effects (i.e., an "over-additive effect").

Testing the presence of a global factor in dyslexia

To evaluate the hypothesis presented above, we set out to systematically test the presence of a global factor in developmental dyslexia. In different studies, because of partially different objectives, we referred to the RAM (e.g., Zoccolotti, De Luca, Judica, & Spinelli, 2008) or the DEM (De Luca, Marinelli, Spinelli, & Zoccolotti, 2017) models. However, as stated above, the information provided by these two models should be seen as complementary rather than alternative.

We illustrate this approach with reference to a study in which we tested words and non-words of different length in children with and without a reading deficit (Martelli, De Luca, Lami, Pizzoli, Pontillo, Spinelli, & Zoccolotti, 2014). The main results are presented in Figure 1a. As it is clear from the graph, the vocal RTs of the children with dyslexia were markedly dependent on stimulus length for both words and non-words; furthermore, they were generally slower across all conditions including in the case of shortest stimuli used (4-letter words and non-words). The experimental effects were generally much smaller in typically developing children but some length effect emerged in the case of non-words.

In Figure 2, data are plotted to detect the presence of global components in the data. In particular, in Figure 2a, data are presented in the form of a so-called Brinley plot, i.e., the condition means for the group of children with dyslexia are plotted as a function of the condition means of the control group. It is very clear that the two sets of data are strongly related; in particular, the differences between the two groups grow linearly as a function of condition difficulty. The coefficient of determination of the linear regression indicates a high level of explanation ($R^2 = 0.88$) and the slope of the regression (3.44) a strong multiplicative difference between the two groups; thus, this slope provides a measure of the global impairment of the dyslexic group.

Figure 1. a) Raw Vocal RTs to words and non-words of different length in children with dyslexia and typically developing children; b) the same data are plotted in the form of individually based z-score values (see text for details in the calculation of z scores). Reproduced from Martelli et al. (2014), with permission from Springer Nature.

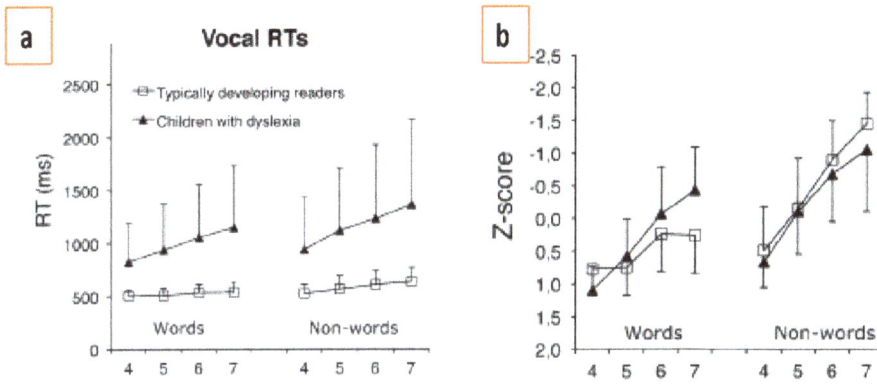

Figure 2b illustrates the relationship between the condition means and the corresponding standard deviations separately in the two groups of children (as advocated by the DEM). It is very clear that, across all conditions, the inter-individual variability linearly grows as a function of condition difficulty. In terms of the presence of a global factor, these data indicate a strong dependency of inter-individual variability on the basic difficulty level of the experimental conditions (with a slope of the linear regression of .87). In keeping with the DEM's predictions, it is of note that, although data for the two groups of children lie in quite different ranges of performance, the basic relationship between means and SDs is essentially the same ($R^2 = 0.99$). Thus, children with dyslexia are much slower than control children but their inter-individual variability in RTs grows by the same factor as it occurs in typically developing children. According to Myerson et al. (2003), this is a critical finding in detecting the presence of a global factor in the data. Therefore, this pattern is consistent with the idea that part of the group differences between children with dyslexia and typically developing children can be ascribed to the presence of a global factor.

Furthermore, according to these authors, this relationship allows decomposing the components of the response due to the sensory-motor compartment and to the cognitive-decisional one. Thus, the intercept on the x-axis of this relationship would represent an estimate of the sensory-motor compartment (a constant value). In this particular case, this amounts to 438 ms. As the same linear regression accounts for the performance of both groups, it follows that the two groups are not different in the sensory-motor compartment but only in the cognitive-decisional one.

We also note that the close relationship between means and SDs across a large range of values represents an important and systematic deviation from the assumption of the homogeneity of variance which is critical for performing parametric analyses such as ANOVAs. Thus, these data pose serious concerns on the standard statistical analyses carried out on raw RT data in comparing children with dyslexia with controls. As these analyses assume homogeneity of variance, they use a single error term to analyze group differences across conditions. However, this will bias results by providing a lenient test for "more difficult" conditions (which are tested against an error term which erroneously includes variability in "simpler" conditions) as well as a too stringent test for simpler conditions. As an effect of this, there is a consistent risk in using standard parametric analyses with raw RTs to artifactually conclude that the group differences in more difficult conditions indeed indicate specific effects.

Figure 2. a) Brinley plot matching the performance of children with dyslexia and typically developing children across the experimental conditions presented in Figure 1; b) SDs are plotted against conditions means on the corresponding conditions; different symbols refer to data of children with dyslexia and typically developing children. Reproduced from Martelli et al. (2014), with permission from Springer Nature.

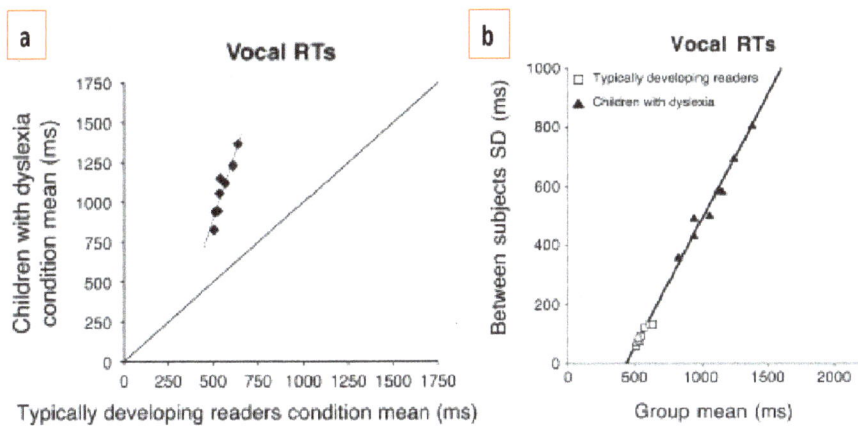

Overall, Figure 2a and 2b give strong indications on the role of global components in the group differences between children with dyslexia and controls. In particular, the two groups are not different in terms of the sensory-motor components of the responses. However, apart from this constant value, the differences between the two groups show a strong multiplicative factor as a function of condition difficulty. Thus, over and above the specific nature of a given experimental condition, the group differences become progressively greater with more difficult conditions and inter-individual variability becomes progressively greater. Note that this pattern is compatible with the basic tenet of both the RAM and DEM models. In particular, both models see the origin of the multiplicative nature of the group differences in the presence of an interaction between an individual general factor (i.e., cognitive speed or rate) and the resources needed to perform the task (difficulty or amount).

Is it possible to test group differences in critical conditions by controlling for the effect of a global factor? The RAM (while not the DEM) provides some possible solutions for responding to this question. In particular, Faust et al. (1999) propose a number of data transformations, such as the use of linear regression analyses or individually-based z-scores. Z-scores indicate an individual participant's performance in a given condition relative to all other conditions based on the individual means of all conditions (therefore, each individual has an average of 0 across conditions and a SD = 1). This transformation re-scales individual performance to a common reference; hence, it allows controlling for global components while it preserves the information regarding individual variability across experimental conditions.

75

The data in Figure 1A have been re-plotted using this data transformation. The transformed data in Figure 1b actually illustrate a significant group x lexicality x length interaction. If one controls for the presence of a global difference between the two groups, the group difference in reading non-words disappears; indeed, both groups are similarly influenced by the length of the target stimulus. By contrast, only children with dyslexia show an influence of length in the case of words while this effect is much reduced in control children.

Overall, the results of this study illustrate several key findings in framing the performance of children with dyslexia within a global factor perspective.

One finding is that children with dyslexia are spared in the sensory-motor compartment and the deficit is specific for the cognitive decisional component of the response. A similar conclusion was reached in a lexical decision study by Zeguers, Snellings, Tijms, Weeda, Tamboer, Bexkens, and Huizenga. (2011) who, based on the diffusion model developed by Ratcliff (1978), observed no difference between children with dyslexia and controls in the non-decision component of RTs. In general, these observations raise some problems for models of developmental dyslexia which posit the locus of the deficit at a sensory level as is in the case of the magnocellular processing hypothesis (Stein, 2001) or at motor level as in case of the idea of a deficit in eye movement programming (e.g., Bucci et al., 2008).

The deficit in children with dyslexia is selective for the cognitive compartment and is of a multiplicative nature such that group differences tend to grow numerically as a function of condition difficulty over and above the specific nature of the experimental conditions involved (over-additivity effect). Thus, there is a risk to interpret group differences in more difficult conditions (such as those involving non-words as compared to words) at face value. By contrast, when the effect of the global factor is controlled for, the putatively specific deficit in reading non-words no longer holds (e.g., Zoccolotti et al., 2008). This conclusion is similar to that of van den Broeck and Geudens (2012) who based their analyses on a more general model, i.e., the state trace analysis. In this approach, the relationship between two different key tasks is examined at different levels of difficulty and used as a reference (or trace) to examine the performance of a group with impaired performance and test whether it represents an actual deviation from expectancy.

In further research, we pursued two general questions connected with these findings. On the one hand, we tried to examine which group differences in psycholinguistic effects survived after controlling for the role of the global factor. On the other hand, we further examined the nature of the global factor itself, i.e., we tried to specify the characteristics of the "rate" component in the genesis of the observed group differences.

Psycholinguistic effects after controlling for the role of the global factor

Evidence on the role of lexical processing in dyslexia comes from a study of the effect of orthographic neighborhood size (N-size) in relationship with word frequency (Marinelli, Traficante, Zoccolotti, & Burani, 2013). As N-size strongly interacts with stimulus length (with many more neighbors for shorter than longer words), the study only examined short words (4- or 5-letter long). Children with dyslexia read low-frequency words with high N-size faster than words that had no neighbors. Notably, this effect was present also in the case in which z-transformed values were examined; thus, this effect could not be accounted for by the presence of over-additivity. The facilitating effect of N-size while reading low-frequency words indicates that children with dyslexia benefitted from lexical activation spreading from dense neighborhoods.

A systematic test of the influence of lexical factors controlling for the possible influence of a global factor was carried out by Paizi, De Luca, Zoccolotti, and Burani (2013) in a study in which, in four different experiments, the role of length, frequency and lexicality in both reading and lexical decision tasks were examined. In both tasks, children with dyslexia showed slower RTs across all conditions. Further, they also showed all effects tested; namely, their RTs were slower for non-words than words, as well as for low-frequency words than for high-frequency words. Notably, the two effects were clearly distinct even though they were tested simultaneously, i.e., by presenting high- and low-frequency words mixed with non-words. Indeed, not only typically developing children, but also children with dyslexia responded faster to low-frequency words than non-words. Furthermore, confirming previous results, RTs of children with dyslexia (but not those of typically developing children) were slower for longer than shorter words.

As expected, data (across the four separate experiments) indicated a close relationship between condition means and SDs, similar to that shown in Figure 2b (the variance accounted for by the linear relation was $R^2 = .93$). Furthermore, plotting data in a Brinley plot showed a linear regression accounting for .94 of variance; response times of children with dyslexia were delayed by more than a factor of 2 compared to typically developing children, i.e., the slope of the Brinley plot was 2.55. These analyses confirmed that global, non-task specific influences were present in the data.

Using a z-score transformation, we set out to detect which of the factors considered (if any) yielded a specific influence after controlling for global influences. Results indicated that children with and without a reading disturbance had a comparable size of the lexicality effect as well as of the frequency effect. Importantly, these findings were confirmed both in the reading and lexical decision tasks.

By contrast, a residual specific effect of length was observed even after controlling for global influences. Again, this finding was present both in the case of the reading and lexical decision tasks. Note that the effect of length was tested only on words; thus, this finding is consistent with the results presented in Figure 1b indicating a selective role of length, after controlling for global components, in the case of words only.

Overall, the findings from Paizi et al. (2013) are in keeping with the idea that lexical activation is present also in children with dyslexia. Remarkably, this finding refers to a language characterized by high orthographic regularity. Indeed, this is contrary to the hypothesis that their reading in these languages is primarily non-lexical (an idea originally referred to as "*orthographic depth hypothesis*"; Frost, Katz, & Bentin, 1987).

In a parallel study, Paizi, De Luca, Burani, and Zoccolotti (2011) further examined the performance in reading words and non-words by comparing a condition in which the two types of stimuli were mixed with to one another and one in which high-frequency words, low-frequency words, and non-words were presented in separate (or pure) blocks. Also in this study, a frequency effect was present in children with dyslexia. However, in this case, the effect was different in the two groups as a function of list context (even after controlling for the presence of a global factor by using z-transformed data). Thus, controls showed a context effect for both high- and low-frequency words; children with dyslexia showed a context effect only for high- but not for low-frequency words. This pattern seems to indicate that lexical organization in children with dyslexia is similar to that of skilled readers but may be underdeveloped: only high-frequency words were dealt with as existing words. In the case of low-frequency words, children with dyslexia were not as flexible as in the case of high-frequency words in switching to lexical reading.

Evidence indicating a possible deficiency for low-frequency items also comes from studies based on an orthographic judgment task (e.g., Angelelli, Marinelli, & Zoccolotti 2010; Marinelli, Angelelli, Notarnicola,& Luzzatti, 2009; Marinelli, Cellini, Zoccolotti, & Angelelli, 2017; note that these investigations were based on accuracy measures; accordingly, the RAN or DEM models could not be applied). In all studies, the children with dyslexia showed selective impairments in detecting phonologically plausible fakes, while their performance was normal when required to judge errors inserted in words with regular orthography. Similar results, pointing to poor lexical reading in dyslexic children of a consistent orthography, have been reported by Bergmann and Wimmer (2008) in a study on Austrian children.

Overall, as stated above (Angelelli et al., 2010; Marinelli et al. 2009, 2017), children with dyslexia showed larger frequency effect respect to controls, with a greater advantage in judging high-frequency stimuli with respect to low-frequency one. This finding highlights that children with dyslexia have spared lexical processing for high frequency stimuli, but they may also have an underdeveloped orthographic lexicon and consequently rely on sublexical processing for low-frequency words.

A separate line of research examined the influence of the morphological structure in the reading of children as a function of the presence/absence of a reading impairment as well as of reading acquisition (Burani, Marcolini, & Stella, 2002).

An original study examined the performance in reading pseudo-words made up of roots and derivational suffixes resulting in nonexistent combinations (e.g., donnista, 'womanist', composed of donn-, 'woman', and -ista, '-ist') and matched pseudo-words with no morphological constituents. Same-age and younger (grades 2/3) control children as well as children with dyslexia named more quickly and accurately the morphological pseudo-words than pseudo-words with no morphological constituents, demonstrating sensitivity to morphological structure (Burani, Marcolini, De Luca, & Zoccolotti, 2008). Younger skilled children were also faster in reading words composed of a root and a derivational suffix (e.g., cassiere, cass-iere; 'cashier'), called complex words, than words with the same length and frequency but without a root (e.g., cammello, 'camel'), called simple words. A similar pattern was observed in sixth graders with dyslexia. By contrast, there was no advantage for complex vs. simple words in typically developing children attending sixth grade who could effectively use whole-word recognition (Burani et al., 2008). A separate study tried to separate the influence of the root and the suffix: results indicated that the facilitation of vocal RTs depended exclusively on roots, while there was no significant effect of suffixes; this pattern held for both children with dyslexia and skilled young readers (Traficante, Marcolini, Luci, Zoccolotti, & Burani, 2011).

In further work, we examined the morphological effect in relationship to word frequency and took into consideration the possible role of global components in the data (Marcolini, Traficante, Zoccolotti, & Burani, 2011). Results showed that word frequency affects the probability of morpheme-based reading in interaction with reading ability. Typically developing children named low- but not high-frequency morphologically complex words faster than simple words. By contrast, the advantage for morphologically complex words was present in poor readers irrespective of word frequency. Notably, the group × morphological type × frequency interaction was significant also in the analysis based on z-transformed values, indicating that it could not be accounted for by an over-additivity effect.

These results indicate that relying on morphemic parsing favors reading fluency in the case of low-frequency words and in children who do not yet fully master whole-word processing (for a review, see Burani, 2010). The availability of preassembled morphemic units favors efficient naming relative to the laborious and slower non-lexical process of segmenting, converting, and reassembling smaller units, necessary for reading new words with no morphological structure.

Overall, lexical information is present in children with dyslexia, as indicated by frequency and lexicality effects, sensitivity to morphological structure and facilitation by orthographic N-size. However, the presence of a length effect in children with dyslexia surviving after controlling for over-additivity, as well as the facilitation in the case of morphologically complex words, point to a reading processing which fails to capture the whole word at a glimpse and requires parsing (no longer necessary in typically developing children of the same age). Indeed, there is some evidence that, while children with dyslexia show some ability to activate lexical information, they may be less able in activating low-frequency entries in the lexicon. Thus, for these stimuli they are inefficient in switching between reading modes (as in passing from a mixed to a pure list context). Further, they show selective difficulties in performing an orthographic judgment test failing to detect phonologically plausible fakes. Thus, in children with dyslexia, entries for low-frequency words may not be as readily available in the orthographic lexicon as in the case of typically developing children.

The nature of the global factor in developmental dyslexia

Results across several different studies confirmed the presence of a multiplicative factor in the data of dyslexic children, as assessed in Brinley plots (De Luca, Burani, Paizi, Spinelli, & Zoccolotti, 2010; De Luca et al., 2017; Di Filippo et al. 2006; Di Filippo & Zoccolotti, 2018 Paizi, et al., 2013; Martelli et al., 2014; Zoccolotti et al., 2008;for a meta-analysis of these data see Zoccolotti, De Luca, Di Filippo, Marinelli, & Spinelli, 2017). Children with different grade level were considered (from third to eighth grade), and investigations included different tasks ranging from reading words of different length and/or frequency, to reading non-words, to simultaneous or sequential matching of strings, and lexical decision tasks. With the exception of a specific effect of length for words which was found in all studies in which this factor was specifically tested on RTs in reading and on lexical decisions (De Luca et al., 2017; Di Filippo et al. 2006; Martelli et al., 2014; Paizi et al., 2013; Zoccolotti et al., 2008), findings consistently confirmed the prediction of a linear relationship between the conditions means of children with dyslexia and those of a control group across many tasks with different levels of difficulty. In particular, the fit of the condition means (R^2) in a Brinley plot match between typically developing children and controls was typically close to 0.90. On the one hand, this linear relationship indicates (as discussed above) an effect of over-additivity, i.e., the presence of larger group differences for more difficult conditions. On the other hand, it points to the idea that a single global factor may explain much of the slowed performance of children with dyslexia; i.e., a single factor is responsible for the lower rate of information processing in children with dyslexia independent from task difficulty. Thus, it seems crucial to understand the nature of such a global factor.

Identifying the scope of the global factor: the concept of domain

To define the nature of the global factor in dyslexia, we performed a number of experiments to detect which conditions contributed to the global factor and which did not conform to this general pattern. In this effort, we took into consideration the concept of "domain", put forward by Myerson et al. (2003) as part of their model.

Myerson et al. (2003) noted that non-task specific influences do not necessarily indicate a general, across the board, effect on all possible tasks. Rather, it is possible to demonstrate that some groups of individuals with "slowed" performance are delayed more on a sub-set of tasks and less in another. One clear example is provided by the effect of aging. Older adults are slow compared to younger adults but the response delay is much more severe in the case of visual-spatial tasks (with slopes ranging from 1.5 to 5.2 depending on the age of the individual) than in the case of verbal tasks (with slopes ranging from 1.2 to 1.5 depending on the age of the individual; Lawrence, Myerson, & Hale., 1998). These results can be accommodated by proposing that aging produces a much greater impairment in a "visual-spatial domain" and a comparatively smaller impairment on a "verbal domain" (Myerson et al., 2003). Note that these impairments are still expressed in a global, not task-specific fashion; thus, the impairment of older individuals across a large variety of tasks in the visual-spatial domain is directly predictable on the basis of a single parameter (the slope of the regression on a Brinley plot match) together with a constant value (the intercept on the x-axis in the plot matching SDs to the corresponding condition means). Myerson et al. (2003) detailed some of the predictions concerning "domains". Critically, while a group of slow individuals (such as older adults) may be more delayed in the tasks mapping a given domain, the relationship (or basic rule determining the global influences) between condition means and SDs remains the same across domains. This has been demonstrated with regard to aging (Myerson et al., 2003). As stated above, older individuals are more delayed in visual-spatial than in verbal tasks, but the relationship in terms of condition means versus SDs between these two sets of data (or domains) remains the same (in particular see Figure 14 in Myerson et al., 2003 which replots data from Hale & Myerson, 1996). Thus, domains represent sets of tasks showing a consistent pattern of results. Actual domains are not given *a priori* but rather they may be empirically determined by establishing the performance of groups of individuals varying for the overall level of information processing across a variety of critical tasks (Myerson et al. (2003) refer to this approach as "*comparative human cognition*"). Based on this framework, in a number of studies we set out to establish the nature of the domain characterizing the non-task specific deficit of children with dyslexia.

Overall, results from different studies pointed to four main findings. First, the global factor includes performance with words, pseudo-words (De Luca et al., 2010, 2017; Marinelli, Angelelli, Di Filippo, & Zoccolotti, 2011; Zoccolotti et al., 2008) and unpronounceable non-words (Marinelli, Traficante, & Zoccolotti, 2014). Thus, the global factor is independent from lexicality. Second, it is not evident if the reading material is a single letter or a bigram; in fact, tasks involving reading of words and non-words yielded a different (greater) slope in a Brinley plot than did tasks involving the reading or matching of letters or bigrams (De Luca et al., 2010). Thus, the global factor explains data for letter strings longer than a bigram. This finding is in keeping with behavioral studies on dyslexia generally indicating spared letter recognition in children with dyslexia (e.g., Bosse et al., 2007; Katz, & Wicklund, 1972; Martelli et al., 2009). Third, if children are tested with acoustically presented materials (i.e., they have to repeat words and pseudo-words or they have to decide on the lexicality of these targets) there is no difference in the rate of information processing between groups (Marinelli et al., 2011); that is, data in the auditory modality do not contribute to the global factor. Thus, the presentation of stimuli in the visual modality is crucial. Fourth, the global factor subtending reading skills does not extend to non-orthographic materials; i.e., when the task is naming pictures, no difference is observed between groups (De Luca et al., 2017; Zoccolotti et al., 2008), implying that the problem in dyslexia is not at the level of phonological access from a visual input *per se*. Overall, based on these results we proposed that the nature of the global factor points to a deficit at the pre-lexical level in the processing of an orthographic string with more than two visually presented characters. Marsh and Hillis (2005) referred to this as "*graphemic description*".

Interestingly, this view is in keeping with neuroimaging results by Dehaene and colleagues (e.g., Dehaene, Cohen, Sigman, & Vinckier, 2005) on the properties of the brain area, referred to as Visual Word Form Area (VWFA). Through visual experience with a given orthography, this area becomes tuned to the more frequent letter sequences (from bigrams to quadrigrams); the VWFA is also connected with other areas responsible for phonological functions. Neuroimaging studies by Blomert and collegues (for a review see Blomert, 2011) point to the presence of specific orthographic-phonological connections (called orthographic-phonological binding). Thus, one possible refinement of the pre-lexical graphemic interpretation is that it is actually concerned with the binding of an orthographic string with phonological trace activation.

Overall, we propose that the nature of the global factor in dyslexia refers to the ability to form and activate pre-lexical processes of "orthographic-phonological binding" upon the visual presentation of an orthographic string (Zoccolotti, De Luca, Di Filippo, Marinelli, & Spinelli, submitted).

Different domains mark the impairment of different groups with slowed level of processing

As an example of how it is possible to use the domain approach to identify the specificity of the dyslexic deficit we illustrate in some detail a study in which we compared the slowness shown by children with dyslexia with respect to peer controls to that shown by older individuals with respect to younger adults (De Luca et al., 2017). Two types of task were considered: reading and picture naming. In particular, naming of objects (pictures) with corresponding names of 4-, 5-, 6- and 7-letters long was investigated; the frequency of the object name was high in half of the trials and low in the other half (separately for each name length). The reading task required reading aloud words and non-words. Words had various lengths, and were matched for word frequency, bigram frequency and initial phoneme across all lengths. Non-words were derived from words by changing one (or two) letter(s) except the first one, and were matched to words for bigram frequency. Some critical results of this study are presented in Figure 3.

Figure 3a reports the Brinley plot for children with dyslexia and controls of the same age with orthographic materials showing a slowing of about 60%, as indicated by the slope (1.6) of the line fitting the reading data points, and no slowing in naming pictures (points lying on the diagonal represent an equal performance between groups). For comparison, Fig 3b shows the Brinley plot comparing groups of older and younger adults. In this case the slowing of older adults (about 25%) was independent from the task, and a single line well fitted both orthographic and pictorial tasks. Older individuals responded slowly compared to younger adults in all considered tasks; thus, their slowing points to a moderate deficit in a more general domain that we can interpret as "verbal". By contrast, the slow response of dyslexic children was highly specific, being limited to orthographic tasks. Thus, in this case, we observe a selective deficit in an "orthographic domain" (dissociated from the performance on pictorial tasks).

Overall, this study provides an example of the possibility to identify in different groups of individuals specific domains of impairment based on the tasks and materials used within the "*comparative human cognition*" perspective proposed by Myerson et al. (2003).

Figure 3. a) Brinley plot matching the performance of children with dyslexia and typically developing children across orthographic and pictorial conditions (reported by different symbols); b) Brinley plot matching the performance of older and younger adults across orthographic and pictorial conditions. Reproduced from De Luca et al. (2017), with permission from Springer Nature.

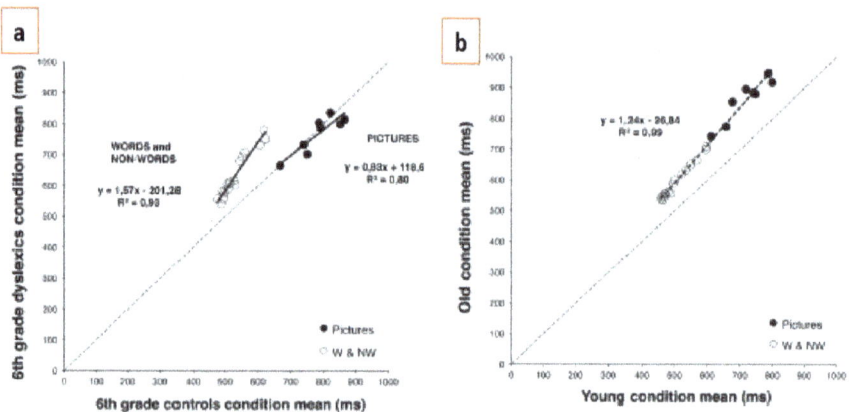

Notably, tasks clustered in different domains for different critical groups. Thus, the impairment of children with dyslexia is well expressed by a deficit limited to an orthographic domain while that of older adults is well expressed as an impairment in a verbal domain (which we know from other studies to be smaller and distinct from a more severe deficit in a visuo-spatial domain; Hale & Myerson, 1996; Lawrence et al., 1998). Overall, the focus on a global approach which controls for possible biases due to over-additivity allows evaluating more reliably the selectivity of the impairment, thus providing important information on the "core deficit" underlying the target behavior.

Synthesis of findings

Much evidence indicates that children with dyslexia are impaired compared to typically developing children in a large variety of reading tasks. Reference to the RAM and the DEM allowed establishing that much of these differences could be accounted for with reference to a global factor. Furthermore, by controlling for the role of global components, it was possible to detect the specific role of some psycholinguistic factors in ways not confounded by the widespread presence of the over-additivity effect.

In particular, based on this approach, we were able to show that lexical processing was present in children with dyslexia, as indicated by the presence of lexicality and frequency effects (Paizi et al., 2013), by the sensitivity to morphological structure (Marcolini et al., 2011) and by the facilitation of N-size in the case of low-frequency words (Marinelli et al., 2013). Still, there was also some evidence that children with dyslexia were inefficient in activating low-frequency entries in the lexicon (e.g., Angelelli et al., 2010; Marinelli et al. 2009, 2017; Paizi et al., 2011). Furthermore, some effects also selectively marked the presence of a reading defect. In particular, the presence of a length effect in children with dyslexia survived after controlling for over-additivity (*e.g.*, Martelli et al., 2014). Furthermore, only children with dyslexia showed facilitation in the case of morphologically complex words as compared to simple words (Burani et al., 2008). These two effects seem to indicate the need for children with dyslexia to parse the orthographic string in order to be able to process it effectively.

Through this approach, we could characterize the nature of the global factor accounting for the performance differences between dyslexic and control children. Examining performance of children with dyslexia across many different experimental conditions it is possible to cancel out the effects due to the level of difficulty of the task and let emerge the non-task specific characteristics of the deficit, as a global factor affecting performances within a given domain. Thus, applying models, such as the RAM or the DEM, provided interesting information on the core deficit of dyslexia. In terms, the same "global" defect was present whenever a string of letters (not a single letter or bigram) was presented, and whether or not it constituted a word; i.e., the same deficit was observed in the case of words, pseudo-words and also unpronounceable non-words. The difficulty did not extend to object naming or responding to stimuli presented in a non-visual modality.

This pattern of findings indicates that the global factor marks the ability to form a pre-lexical representation of a visually presented orthographic string and that this may indeed represent the core process of reading competence. We have already noted that this proposal is similar to the idea of a "*graphemic description*", put forward by Marsh and Hillis (2005) as well as, at the neural level, with the characteristics of the VWFA as proposed Dehaene and colleagues (e.g., Dehaene et al., 2005). At least in the case of languages with high orthographic regularity, developmental dyslexia would be characterized by a deficit in forming pre-lexical representations of visually presented orthographic strings.

Conclusions

As we have seen at the beginning of this chapter, as a group, children with dyslexia are impaired in a variety of perceptual, attentional, cognitive and motor tasks and this has led to the development of a large number of hypotheses on the reading disturbance. How can the proposal of a deficit in forming pre-lexical representations be reconciled with this large multitude of deficits?

This problem can be tackled from different (but possibly not mutually exclusive) perspectives.

First, one way to frame the various group differences and related theories (briefly outlined at the beginning of this chapter) is to consider that they deal with reading at different levels of processing. Since the seminal work of Morton and Frith (1995), it is well established that developmental disorders need to be seen in the their full biological, cognitive and behavioral expression. Similarly, in his "*multiple deficit model*", Pennington (2006) distinguishes between the level of complex behavioral disorders (which is the one at which disorders such as dyslexia or other learning disorders are examined from a clinical/diagnostic standpoint) from the level of cognitive processes, the level of neural systems and that of etiological risk and protective factors. If one examines the current views on dyslexia, it is clear that different theoretical approaches focus on different levels of processing. Thus, views such as the cerebellar hypotheses of dyslexia (Nicolson et al., 2001) or the magnocellular theory (e.g., Stein, 2001) largely focus on the neural level, though they also comment upon the cognitive components of the disturbance. By contrast, other theories, such as the core phonological hypothesis of dyslexia (e.g., Stanovich, 1988) or the visual attention span deficit hypothesis (e.g., Bosse et al., 2007) focus on the cognitive processes. While levels are certainly interconnected, hypotheses working at different levels of processing cannot be directly compared. As it regards our present proposal, a deficit in forming pre-lexical representations deals with the cognitive level. Speculatively, we have quoted neural evidence which appears compatible with our hypothesis (Dehaene et al., 2005); however, the two levels should remain distinct.

Second, focusing at the cognitive level, one issue concerns whether deficits associated with dyslexia do represent causes or, rather, should be seen as consequences of the reading disturbance itself. This possibility has been raised in the case of various cognitive factors and even in the case of sensory deficits (Goswami, 2015). For example, it has been known for a while that children develop awareness of words as being made of a sequence of phones following formal training to read, not as a spontaneous consequence of cognitive development (Morais, Gary, Alegria, & Bertelson, 1979). Additional information on this point comes from studies which compared children who started school later than expected with children with the same age but differing for school experience. Results indicated that meta-phonological skills closely varied as an effect of school entrance, i.e., they were higher in children entering school earlier than in in peers who entered school later (e.g., Cunningham & Carroll 2011). Thus, deficient meta-phonological skills can be most parsimoniously seen as an effect of the reading disturbance rather than a cause. A similar argument has been recently put forward also in the case of rapid automatized naming (Peterson, Arnett, Pennington, Byrne,

Samuelsson, & Olson, 2018).

Thirdly, one important question on cognitive factors concerns the distinction between proximal and distal processes (Coltheart, 2015). Proximal processes make explicit the cognitive antecedents of reading. In this sense, the ability to apply rules of grapheme-to-phoneme conversion would represent a proximal process of reading (Coltheart, 2015). We think that focusing on proximal processes of reading, as done in the present approach, is a convenient first step in the analysis of dyslexia. Distal processes (such as short-term memory, memory retrieval, attention, etc.) influence the reading process not directly, but through the action of proximal processes (Coltheart, 2015). One can reasonably move to study distal processing as a second step; in this effort, it would be crucial to make explicit the relationship between distal and proximal processes (a specification which is rarely spelled out).

How does the approach described in the present paper fit in this complex picture? We started from the observation that raw data do not allow discriminating different sources of individual variability because performance in a given reading condition depends on both task's (such as the general level of difficulty) and subject's characteristics (such as the general level of difficulty and the individual capacity to respond quickly to a class of speeded tasks). Thus, relying only on raw data in single tasks does not allow discriminating different sources of individual variability. This makes it difficult to identify a critical task condition that would capture the specificity of the reading deficit, as it has been proposed for the non-word reading task (e.g., Rack et al., 1992).

Models such as the RAM or DEM provide explicit predictions on the sources of variance determining individual variability in speeded tasks, and can be extended to reading. Thus, it is possible to quantify the reading deficit in ways that are independent of the difficulty level of a given task and point to a general, task-independent impairment. This was instrumental in pinpointing the critical factor affecting the reading deficit, i.e., a deficit in forming a pre-lexical representation of a visually presented orthographic string. In this vein, this defect can be seen as the key competence which determines the reading skill over and above the influence of task-specific influences. However, some studies provided evidence for a second deficit in children with dyslexia. Namely, lexical organization in these children is similar to that of skilled readers but may be underdeveloped; in particular, children with dyslexia are inefficient in activating low-frequency entries in the lexicon (e.g., Angelelli et al., 2010; Marinelli et al., 2017, 2019; Paizi et al., 2011). At present, the source of this latter defect is not entirely clear. An intriguing possibility is that the deficit in activating representations for low-frequency words is indirectly associated to the key deficit in forming graphemic representations. Thus, an early pre-lexical deficit might alter later stages of processing dampening the ability to establish representations in the orthographic lexicon. A longitudinal study of reading skills seems required to test this possibility.

References

Angelelli P., Marinelli C. V., & Zoccolotti P. (2010). Single or dual representation for reading and spelling? A study on Italian dyslexic and dysgraphic children. *Cognitive Neuropsychology,* 27, 305- 333.

Barca, L., Burani, C., & Arduino, L. S. (2002). Word naming times and psycholinguistic norms for Italian nouns. *Behavior Research Methods, Instruments, & Comp*uters, *34,* 424-434.

Barca, L., Burani, C., Di Filippo, G. & Zoccolotti, P. (2006). Italian developmental dyslexic and proficient readers: Where are the differences? *Brain and Language, 98,* 347-351.

Barca, L., Ellis, A. W., & Burani, C. (2007). Context-sensitive rules and word naming in Italian children. *Reading and Writing,* 20, 495-509.

Barton, J. J., Hanif, H. M., Eklinder Björnström, L., & Hills, C. (2014). The word-length effect in reading: a review. *Cognitive Neuropsychology, 31,* 378-412.

Bergmann, J., & Wimmer, H. (2008). A dual-route perspective on poor reading in a regular orthography: Evidence from phonological and orthographic lexical decisions. *Cognitive Neuropsychology, 25,* 653-676.

Blomert, L. (2011). The neural signature of orthographic–phonological binding in successful and failing reading development. *Neuroimage, 57,* 695-703.

Bosse, M. L., Tainturier, M. J., & Valdois, S. (2007). Developmental dyslexia: The visual attention span deficit hypothesis. *Cognition, 104,* 198-230.

Bruck, M. (1992). Persistence of dyslexics' phonological awareness deficits. *Developmental Psychology, 28,* 874-886.

Bryant, P., MacLean, M., & Bradley, L. (1990). Rhyme, language, and children's reading. *Applied Psycholinguistics, 11,* 237-252.

Bucci, M. P., Brémond-Gignac, D., & Kapoula, Z. (2008). Latency of saccades and vergence eye movements in dyslexic children. *Experimental Brain Research, 188,* 1-12.

Burani, C., Marcolini, S., & Stella, G. (2002). How early does morpholexical reading develop in readers of a shallow orthography?. *Brain and Language, 81,* 568-586.

Burani, C., Barca, L., & Ellis, A. W. (2006). Orthographic complexity and word naming in Italian: Some words are more transparent than others. *Psychonomic Bulletin & Review, 13,* 346-352.

Burani, C., Marcolini, S., De Luca, M. & Zoccolotti, P. (2008). Morpheme-based reading aloud: Evidence from dyslexic and skilled Italian readers. *Cognition, 108,* 243-262.

Burani, C. (2010). Word morphology enhances reading fluency in children with developmental dyslexia. *Lingue e linguaggio, 9,* 177-198.

Castles, A. & Coltheart, M. (1993). Varieties of developmental dyslexia. *Cognition, 47,* 149-180.

Cerella, J. (1985). Information processing rates in the elderly. *Psychological Bulletin, 98,* 67.

Coltheart, M., Rastle, K., Perry, C., Langdon, R., & Ziegler, J. (2001). DRC: a dual route cascaded model of visual word recognition and reading aloud. *Psychological Review, 108,* 204-256.

Coltheart, M. (2015). What kinds of things cause children's reading difficulties? *Australian Journal of Learning Difficulties, 20,* 103-112.

Cunningham, A. J., & Carroll, J. M. (2011). Reading-related skills in earlier- and later-schooled children. *Scientific Studies of Reading, 15,* 244-266.

De Luca, M., Di Pace, E., Judica, A., Spinelli, D. & Zoccolotti, P., (1999). Eye movement patterns in linguistic and non-linguistic tasks in developmental surface dyslexia, *Neuropsychologia, 37,* 1407-1420.

De Luca, M., Borrelli, M., Judica, A., Spinelli, D., & Zoccolotti, P. (2002). Reading words and pseudowords: An eye movement study of developmental dyslexia. *Brain and Language, 80,* 617-626.

De Luca, M., Barca, L., Burani, C. & Zoccolotti, P. (2008). The effect of word length and other sublexical, lexical and semantic variables on developmental reading deficit. *Cognitive and Behavioral Neurology, 21,* 227-235.

De Luca, M., Burani, C., Paizi, D., Spinelli, D. & Zoccolotti, P. (2010). Letter and letter-string processing in developmental dyslexia. *Cortex, 46,* 1272- 1283.

De Luca, M., Marinelli, C.V., Spinelli, D., & Zoccolotti, P. (2017). Slowing in reading and picture naming: the effects of aging and developmental dyslexia. *Experimental Brain Research, 235,* 3093-3109.

Dehaene, S., Cohen, L., Sigman, M., & Vinckier, F. (2005). The neural code for written words: a proposal. *Trends in Cognitive Sciences, 9,* 335-341.

Denckla, M. B., & Rudel, R. G. (1976). Rapid 'automatized'naming (RAN): Dyslexia differentiated from other learning disabilities. *Neuropsychologia, 14,* 471-479.

Di Filippo, G., De Luca, M., Judica, A. Spinelli, D. & Zoccolotti, P. (2006). Lexicality and stimulus length effects in Italian dyslexics: role of over- additivity effect. *Child Neuropsychology, 12,* 141-149.

Di Filippo, G. & Zoccolotti, P. (2018). Isolating global components in developmental dyscalculia and dyslexia. *Frontiers in Psychology, 9*:171.

Facoetti, A., Trussardi, A. N., Ruffino, M., Lorusso, M. L., Cattaneo, C., Galli, R., ... & Zorzi, M. (2010). Multisensory spatial attention deficits are predictive of phonological decoding skills in developmental dyslexia. *Journal of Cognitive Neuroscience, 22,* 1011-1025.

Farmer, M. E., & Klein, R. M. (1995). The evidence for a temporal processing deficit linked to dyslexia: A review. *Psychonomic Bulletin & Review, 2,* 460-493.

Faust, M. E., Balota, D. A., Spieler, D. H., & Ferraro, F. R. (1999). Individual differences in information-processing rate and amount: Implications for group differences in response latency. *Psychological Bulletin, 125,* 777-799.

Fawcett, A. J., & Nicolson, R. I. (1995). Persistent deficits in motor skill of children with dyslexia. *Journal of Motor Behavior, 27,* 235-240.

Frost, R., Katz, L., & Bentin, S. (1987). Strategies for visual word recognition and orthographical depth: a multilingual comparison. *Journal of Experimental Psychology: Human Perception and Performance, 13,* 104-115.

Goswami, U. (2011). A temporal sampling framework for developmental dyslexia. *Trends in Cognitive Sciences, 15,* 3-10.

Goswami, U. (2015). Sensory theories of developmental dyslexia: three challenges for research. *Nature Reviews Neuroscience, 16,* 43-54.

Hale, S., & Myerson, J. (1996). Experimental evidence for di erential slowing in the lexical and nonlexical domains. *Aging Neuropsychology and Cognition, 3,* 154-165.

Hämäläinen, J. A., Salminen, H. K., & Leppänen, P. H. (2013). Basic auditory processing deficits in dyslexia: Systematic review of the behavioral and event-related potential/field evidence. *Journal of learning disabilities, 46,* 413-427.

Hermann, J.A., Matyas, T. & Pratt, R. (2006) Meta-analysis of the nonword reading deficit in specific reading disorder. *Dyslexia, 12,* 195-221.

Kapoula, Z., & Bucci, M. P. (2007). Postural control in dyslexic and non-dyslexic children. *Journal of Neurology, 254,* 1174-1183.

Katz, L., & Wicklund, D. (1972). Letter scanning rate for good and poor readers in grades two to six. *Journal of Educational Psychology, 63,* 363-367.

Lawrence, B., Myerson, J., & Hale, S. (1998). Differential decline of verbal and visuospatial processing speed across the adult life span. *Aging Neuropsychology and Cognition, 5,* 129-146

Lieberman, P., Meskill, R. H., Chatillon, M., & Schupack, H. (1985). Phonetic speech perception deficits in dyslexia. *Journal of Speech, Language, and Hearing Research, 28,* 480-486.

Marcolini, S., Traficante, D., Zoccolotti, C., & Burani, C. (2011). Word frequency modulates morpheme-based reading in poor and skilled Italian readers. *Applied Psycholinguistics, 32,* 513–532. doi:10.1017/Sù 0142716411000191

Marinelli, C. V., Angelelli, P., Notarnicola, A., & Luzzatti, C. (2009). Do Italian dyslexic children use the lexical reading route efficiently? An orthographic judgment task. *Reading and Writing, 22,* 333-351.

Marinelli, C.V., Angelelli, P., Di Filippo, G. & Zoccolotti, P. (2011). Is developmental dyslexia modality specific? A visual-acoustic comparison on Italian dyslexics. *Neuropsychologia, 49,* 1718-1729.

Marinelli, C.V., Traficante, D., Zoccolotti, P. & Burani, C. (2013). Orthographic neighborhood-size effects on the reading aloud of Italian children with and without dyslexia. *Scientific Studies of Reading, 17,* 333-349.

Marinelli, C.V., Traficante, D., & Zoccolotti, P. (2014). Does pronounceability modulate the letter string deficit of children with dyslexia? Evidence from lexical decision task. *Frontiers in Psychology, 5:* 1353.

Marinelli C.V., Cellini P., Zoccolotti P., & Angelelli P. (2017). Whole word processing and sensitivity to written regularity in a consistent orthography: a reading and spelling longitudinal study on dyslexic and typically developing children. *Cognitive Neuropsychology, 34,* 163-186.

Marsh, E. B., & Hillis, A. E. (2005). Cognitive and neural mechanisms underlying reading and naming: evidence from letter-by-letter reading and optic aphasia. *Neurocase, 11,* 325-337.

Martelli, M., Di Filippo, G., Spinelli, D., & Zoccolotti, P. (2009). Crowding, reading, and developmental dyslexia. *Journal of Vision, 9,* 14-14.

Martelli, M., De Luca, M., Lami, L., Pizzoli, C., Pontillo, M., Spinelli D., & Zoccolotti, P. (2014). Bridging the gap between different measures of the reading speed deficit in developmental dyslexia. *Experimental Brain Research, 232,* 237-252.

Melby-Lervåg, M., Lyster, S. A. H., & Hulme, C. (2012). Phonological skills and their role in learning to read: a meta-analytic review. *Psychological Bulletin, 138,* 322-352.

Mervis, C. B., & Klein-Tasman, B. P. (2004). Methodological issues in group-matching designs: α levels for control variable comparisons and measurement characteristics of control and target variables. *Journal of Autism and Developmental Disorders, 34,* 7-17.

Morais, J. Gary, L., Alegria, J., & Bertelson, P. (1979). Does awareness of speech as a sequence of phones arise spontaneously? *Cognition, 7,* 323-331.

Morton, J., & Frith, U. (1995). *Causal modeling: a structural approach to developmental psychopathology.* In D. Cicchetti & D. J. Cohen (Eds.), Developmental psychopathology (Vol. 1, pp. 357-390). New York: John Wiley & Sons.

Myerson, J., Hale, S., Zheng, Y., Jenkins, L., & Widaman, K. F. (2003). The difference engine: A model of diversity in speeded cognition. *Psychonomic Bulletin & Review, 10,* 262-288.

Nicolson, R. I., Fawcett, A. J., & Dean, P. (2001). Developmental dyslexia: the cerebellar deficit hypothesis. *Trends in Neurosciences, 24,* 508-511.

Paizi, D., De Luca, M., Burani, C. & Zoccolotti, P. (2011). List context manipulation reveals orthographic deficits in Italian readers with developmental dyslexia. *Child Neuropsychology, 17,* 459-482.

Paizi, D., De Luca, M., Zoccolotti, P., & Burani, C. (2013). A comprehensive evaluation of lexical reading in Italian developmental dyslexics. *Journal of Research in Reading, 36,* 303-329.

Pennington, B. F. (2006). From single to multiple deficit models of developmental disorders. *Cognition, 101,* 385-413.

Peterson, R. L., Pennington, B. F., & Olson, R. K. (2013). Subtypes of developmental dyslexia: testing the predictions of the dual-route and connectionist frameworks. *Cognition, 126,* 20-38.

Peterson, R. L., Arnett, A. B., Pennington, B. F., Byrne, B., Samuelsson, S., & Olson, R. K. (2018). Literacy acquisition influences children's rapid automatized naming. *Developmental science, 21,* e12589.

Pozzo, T., Vernet, P., Creuzot-Garcher, C., Robichon, F., Bron, A., & Quercia, P. (2006). Static postural control in children with developmental dyslexia. *Neuroscience Letters, 403,* 211-215.

Rack, J. P., Snowling, M. J., & Olson, R. K. (1992). The nonword reading deficit in developmental dyslexia: a review. *Reading Research Quarterly, 27,* 29-53.

Ramus, F., & Ahissar, M. (2012). Developmental dyslexia: The difficulties of interpreting poor performance, and the importance of normal performance. *Cognitive Neuropsychology, 29,* 104-122.

Ratcliff, R. (1978). A theory of memory retrieval. *Psychological Review, 85,* 59-108.

Ratcliff, R., Thapar, A., Gomez, P., & McKoon, G. (2004). A diffusion model analysis of the effects of aging in the lexical-decision task. *Psychology and Aging, 19,* 278-289.

Raymond, J. E., & Sorensen, R. E. (1998). Visual motion perception in children with dyslexia: Normal detection but abnormal integration. *Visual Cognition, 5,* 389-404.

Share, D. L. (2008). On the Anglocentricities of current reading research and practice: the perils of overreliance on an" outlier" orthography. *Psychological Bulletin, 134,* 584-615.

Spinelli, D. De Luca, M., Di Filippo, G., Mancini, M. Martelli M. & Zoccolotti, P. (2005). Length effect in word naming latencies: role of reading experience and reading deficit. *Developmental Neuropsychology, 27,* 217-235.

Stein, J. (2001). The magnocellular theory of developmental dyslexia. *Dyslexia, 7,* 12-36.

Stanovich, K. E., & Cunningham, A. E. (1992). Studying the consequences of literacy within a literate society: The cognitive correlates of print exposure. *Memory & Cognition, 20,* 51–68.

Stanovich, K. E., Siegel, L. S., & Gottardo, A. (1997). Converging evidence for phonological and surface subtypes of reading disability. *Journal of Educational Psychology, 89,* 114.

Stanovich, K. E. (1988). Explaining the differences between the dyslexic and the garden variety poor readers: The phonological-core variable-difference model. *Journal of Learning Disabilities, 21,* 590-612.

Swan, D., & Goswami, U. (1997). Phonological awareness deficits in developmental dyslexia and the phonological representations hypothesis. *Journal of*

Experimental Child Psychology, 66, 18-41.

Swanson, H. L., Zheng, X., & Jerman, O. (2009). Working memory, short-term memory, and reading disabilities: A selective meta-analysis of the literature. *Journal of Learning Disabilities, 42*, 260-287.

Tallal, P. (1980). Auditory temporal perception, phonics, and reading disabilities in children. *Brain and Language, 9*, 182-198.

Traficante, D., Marcolini, S., Luci, A., Zoccolotti, P. & Burani, C. (2011). How do root and suffix influence reading of morphological pseudowords: A study of young Italian dyslexic children. *Language and Cognitive Processes, 26*, 777-793.

van den Boer, M., de Jong, P. F., & Haentjens-van Meeteren, M. M. (2013). Modeling the length effect: Specifying the relation with visual and phonological correlates of reading. *Scientific Studies of Reading, 17*, 243-256.

van den Broeck, W. & Geudens, A. (2012). Old and new ways to study characteristics of reading disability. The case of the nonword reading deficit. *Cognitive Psychology, 65*, 414-456.

van Ijzendoorn, M. H., & Bus, A. G. (1994). Meta-analytic confirmation of the nonword reading deficit in developmental dyslexia. *Reading Research Quarterly, 29*, 267-275.

Varvara, P., Varuzza, C., Padovano Sorrentino, A. C., Vicari, S., & Menghini, D. (2014). Executive functions in developmental dyslexia. *Frontiers in Human Neuroscience, 8*:120.

Vidyasagar, T. R., & Pammer, K. (2010). Dyslexia: a deficit in visuo-spatial attention, not in phonological processing. *Trends in Cognitive Sciences, 14*, 57-63.

Wagenmakers, E. J., & Brown, S. (2007). On the linear relation between the mean and the standard deviation of a response time distribution. *Psychological Review, 114*, 830-841.

Verhaeghen, P., & Cerella, J. (2002). Aging, executive control, and attention: a review of meta-analyses. *Neuroscience and Biobehavioral Reviews, 26*, 849-857.

Wimmer, H. (1993). Characteristics of developmental dyslexia in a regular writing system. *Applied Psycholinguistics, 14*, 1-33.

Zeguers, M. H., Snellings, P., Tijms, J., Weeda, W. D., Tamboer, P., Bexkens, A., & Huizenga, H. M. (2011) Specifying theories of developmental dyslexia: a diffusion model analysis of word recognition. *Developmental Science, 14*, 1340-1354.

Ziegler, J. C., Perry, C., Ma-Wyatt, A., Ladner, D., & Schulte-Körne, G. (2003). Developmental dyslexia in different languages: Language-specific or universal?. *Journal of Experimental Child Psychology, 86*, 169-193.

Zoccolotti, P., De Luca, M., Di Pace, E., Judica, A., Orlandi, M., & Spinelli, D. (1999). Markers of developmental surface dyslexia in a language (Italian) with high grapheme-phoneme correspondence. *Applied Psycholinguistics, 20*, 191-216.

Zoccolotti, P., De Luca, M., Gasperini, F., Judica, A. & Spinelli, D. (2005). Word length effect in early reading and in developmental dyslexia. *Brain and Language, 93,* 369-373.

Zoccolotti, P., De Luca, M., Judica, A. & Spinelli, D. (2008). Isolating global and specific factors in developmental dyslexia: a study based on the rate and amount model (RAM). *Experimental Brain Research, 186,* 551-560

Zoccolotti, P., De Luca, M., Di Filippo, G., Marinelli, C. V., & Spinelli, D. (submitted). Toward a proximal cognitive model of co-morbidity: Predicting individual differences in reading, writing and maths. *Cognitive Neuropsychology.*

Zoccolotti, P., De Luca, M., Di Filippo, G., Marinelli, C. V., & Spinelli, D. (2017). Reading and lexical decision tasks generate different patterns of individual variability as a function of condition difficulty. *Psychonomic Bulletin and Review, 25,* 1161-1169.

Looking at ancillary systems for verb recovery: evidence from direct current stimulation (DCS)

Francesca Pisano[1], P. Marangolo[1,2]*

[1]IRCCS, Fondazione Santa Lucia, Rome, Italy

[2]Università Federico II, Naples, Italy

*paola.marangolo@gmail.com

Abstract

Over the last years, different studies have suggested that the language function is not restricted into the classical language areas, but it involves regions which have never been considered before. Coherently, very recently, it has been shown that the modulation of neural structures through direct current stimulation (DCS) is also effective when applied over the motor cortex, the cerebellum and the spinal cord. This evidence points to the opportunity to consider the sensorimotor system and its connected structures as part of language recovery. Accordingly, it has been found that DCS over these regions facilitates the retrieval of words associated to motor schemata, such as action verbs.

We will review a series of studies aimed to assess the impact of DCS for verb recovery. In previous experiments, we show that DCS applied over the left Broca's area combined with a naming treatment results efficacious for verb improvement in persons with chronic aphasia. In two very recent protocols, we report these positive effects also after DCS of the cerebellum and the spinal cord. Since verbs play a crucial role for sentence construction and speech fluency which are most often impaired in the aphasic population, we believe that these results have important clinical implications. Indeed, they address the possibility that different nervous structures might act as ancillary systems to support verb recovery.

Since the early nineteenth century, single-cases and anatomo-correlative studies on patients with brain injuries have reported how the language faculty is organized in the human brain. (Broca 1861, Wernicke 1874, Lichtheim, 1885). Antiquitus, most models of language representation have suggested that the different language components are localized in specific areas of the left hemisphere. Thus, single case studies were reported whose lesions to the inferior frontal region (i.e. Broca's area) selectively damaged speech production, while lesions to the posterior superior temporal region (i.e. "Wernicke's area") impaired auditory speech comprehension. In particular, the behavioural and neural processes underpinning different word classes, specifically nouns and verbs, have been a long-standing area of interest in psycholinguistic, neuropsychology and aphasiology research. (Alyahyaa et al., 2018).

Nouns vs. Verbs Dissociation

Evidence from neuropsychological studies

Given that several single left brain-damaged case studies documented a double dissociation between nouns and verbs (Miceli et al., 1988; Caramazza et al., 1991; Laiacona et al., 2004; Tomasino et al., 2018), the researchers hypothesized that the cerebral systems for the recovery of these classes of words were segregated in different areas of the brain. Miceli and colleagues (1988) were the first to report a correlation between word-class specific impairments and lesions in different brain areas. In their study, participants with selective deficits for nouns were mostly affected by lesions in the left temporal areas, whereas participants with a predominant verb impairment showed the involvement of the left frontal cortex. The view that the frontal and the temporal areas were differently involved in processing verbs and nouns, respectively, was further strengthened by Daniele and colleagues (1994). The authors compared the performance of two patients with frontal lobe atrophy and impairment in verb processing with the outcome of a third patient with temporal lobe atrophy and impairment in noun processing.

The anatomical distinction between the two categories is not surprising as nouns and verbs differ in terms of grammatical/syntactic properties as well as in their different amount of sensory/functional features. Nouns usually refer to concrete entities (e.g. objects, tools, animals) intrinsically denoted by a rich set of perceptual features. As suggested by Tranel and colleagues (2001), the temporal lobe belongs to the network that processes perceptual information by mapping the physical features that are crucial for the encoding of concrete entities as well as their associated linguistic knowledge. Therefore, when the processing of a stimulus requires detailed analysis of its perceptual features, the temporal lobe emerges and, subsequently, a lesion to this region might cause an impairment in the retrieval of nouns.

In contrast, verbs typically refer to actions and, thus, are generally denoted by sensorimotor features. The frontal cortex is known to play a crucial role in both the encoding and the production of actions and it is considered the basis for the recognition of meaningful actions (Cappa et al., 2002). Therefore, the link between the frontal cortex and verbs could be at least in part due to the activation of action representation (Parsons et al., 1995; Shapiro et al., 2006).

Evidence from transcranial direct current stimulation studies on verb naming (tDCS)

Since last 10 years, non- invasive brain stimulation devices and, in particular, transcranial direct current stimulation [tDCS] have been introduced in order to implement protocols for the recovery of language (see Marangolo, 2017 for a review). During tDCS, weak polarizing direct currents are delivered to the cortex via two electrodes placed on the scalp. The nature of the effect depends on the polarity of the current. Generally, the anode increases cortical excitability when applied over the region of interest with the cathode above the contralateral orbit or above the shoulder (as the reference electrode), whereas the cathode decreases it, limiting the resting membrane potential. These effects may last for minutes to hours depending on the intensity, polarity and duration of stimulation and they are generally compared with a placebo condition (the so-called "sham" condition) in which the stimulator is turned off after 30 sec (Nitsche and Paulus, 2011).

Following the hypothesis of a segregation of verbs and nouns in different brain regions (Miceli et al., 1994; Tranel et al., 2001), Fiori et al. (2013) investigated whether tDCS, over the left frontal and/or the left temporal regions coupled with an intensive language treatment, would differently improve noun and verb recovery in a group of chronic aphasics. In this study seven patients with post-stroke aphasia were involved in two intensive language treatments for their noun and verb retrieval deficits. During each training, each subject was treated with tDCS (20min, 1mA) over the left hemisphere in three different conditions: anodal tDCS over the left temporal area, anodal tDCS over the left frontal area and sham stimulation, while they performed the naming tasks. Unlike the other two conditions, results showed a significant greater improvement in noun naming after stimulation over the temporal area, while verb naming recovered significantly better after stimulation of the frontal area. Thus, these data confirmed the existence of a segregation of nouns and verbs in different parts of the brain. These results were further confirmed in a subsequent work by Marangolo and colleagues (2013) who found that tDCS combined with a verb naming treatment was efficacious for the recovery of verbs only when delivered over the Broca's region.

However, in more recent years, the hypothesis of a functional segregation for the different classes of words has been questioned (Bifonsky et al., 2006; Siegel et al., 2017; Ulm, Copland & Marcus Meinzer, 2018). Indeed, several neuroimaging studies pointed to a more widely distributed fronto-temporal network involved in nouns and verbs processing with overlapping areas among the two word categories (Tranel et al., 2001; Li et al., 2004; Aggujaro et al., 2006; Luzzatti et al., 2006; Piras & Marangolo, 2007; Siri et al., 2008; Kemmerer et al., 2010a; Kemmerer et al., 2010b). Accordingly, these studies have emphasized that verb difficulties might arise also from damage beyond the left frontal lobe, such as the temporal and the parietal cortex, the basal ganglia and the insula (Tranel et al., 2001; Li et al., 2004; Aggujaro et al., 2006; Luzzatti et al., 2006; Siri et al., 2008; Kemmerer et al., 2010a; Kemmerer et al., 2010b; Crepaldi et al., 2013; Kemmerer et al., 2014; Alyahya et al., 2018). In line with this view, a more interactive view for language processing emerged supported by the "Embodied Cognition" theory.

The Embodied Cognition Theory

Following the hypothesis that language processing might involve "less classical" cortical areas, the Embodied Cognition theory posits that the motor and the language system mutually interact, especially when the words to be processed carry sensorimotor features, such as action verbs (Buccino et al., 2004; Gazzola et al., 2007; Rizzolatti et al., 2009; Casile et al., 2011; Michel et al., 2013). Indeed, it has been demonstrated that when people listen to linguistic descriptions containing verbs, their somatosensory, motor and premotor areas respond by activating a neural population that reflects the corresponding action (Pulvermüller et al., 2005; Binkofski & Buccino 2006; for review see Willems & Hagoort 2007; Fischer & Zwaan 2008). In particular, in response to listening to hand (i.e. to greet), mouth (i.e. to bite) and foot (i.e. to kick) actions, metabolic activity increases in the corresponding motor areas, reflecting a somatotopic organization (Buccino et al., 2001). With regard to verb production, behavioural and neuromodulation studies in healthy subjects have repeatedly shown that the execution of gestures corresponding to actions (i.e. to greet) influence the production of the corresponding word (i.e. to greet) (Bernardis & Gentilucci 2006; Gentilucci & Dalla Volta 2008a; Gentilucci & Dalla Volta 2008b).

Thus, in more recent years, some researchers have questioned whether the motor system might act as an auxiliary support for verb recovery in the aphasic population. Coherently with the above results, it was found that the use of gestures congruent with the corresponding action, influences the recovery of the corresponding verbs (Marangolo et al., 2010; Marangolo et al., 2012; Bonifazi et al., 2013).

Motor cortex stimulation for verb recovery

Given the above reported interaction between the motor and the language system, some studies have investigated, in the healthy and brain damaged populations, whether the direct involvement of the left motor cortex through tDCS would influence action verb processing (Liuzzi et al., 2010; Meinzer et al., 2016; Branscheidt et al., 2017). Liuzzi and colleagues (2010) explored the effects of anodal, cathodal and sham tDCS over the left motor cortex combined with a novel action word vocabulary in a group of 30 healthy volunteers. Results revealed a specific inhibitory effect of cathodal tDCS in the subject's ability to associate actions with novel words with respect to the sham condition. Branscheidt and colleagues (2017) investigated whether the motor cortex is involved in accessing specific lexical-semantic information (such as object vs. action words) in a group of 16 post-stroke aphasics. Participants were asked to perform a lexical decision task judging whether the presented stimuli were words or pseudowords. In all subjects, anodal tDCS improved accuracy in lexical decision, especially for words with action-related content and for pseudowords with an "action-like" ending but not for words with object-related content and pseudowords with "object-like" characteristics. Thus, the hypothesis was advanced that motor cortex stimulation has strengthen content-specific word-to-semantic concept associations only for words, such as action verbs, associated to motor schemata.

Therefore, given the above reported involvement of the motor cortex in verb retrieval, the hypothesis was advanced that other sensorimotor regions connected to the motor cortex would result effective, if modulated through DCS, for language recovery.

Cerebellar stimulation for verb recovery

The hypothesis of a possible involvement of the cerebellum in language processing is not recent. Indeed, in these last years, several linguistic disorders after acquired cerebellar lesions have been documented (De Smet et al., 2007), such as impaired verbal fluency (Schmahmann & Sherman, 1998; Leggio et al., 2000; Richter et al., 2004; Stoodley & Schmahmann, 2009), agrammatism (Schmahmann & Sherman, 1998) and naming difficulties (Schmahmann & Sherman, 1998; Gasparini et al., 1999; Fabbro et al., 2000). The frequent co-occurrence of a right cerebellar lesion and aphasia led some authors to hypothesize the existence of a "lateralized linguistic cerebellum" (Mariën et al., 1996, 2000, 2014). According to Mariën et al. (1996, 2000, 2014), the aphasic disorder reflects a "diaschisis" phenomenon whereby the damage of the right cerebellum causes a hypofunction of the left frontal cortical areas, "home" of our language representation (Mariën et al., 1996, 2000, 2014). The cerebellum would, thus, have a role in language processing but only through its connections with the left frontal language areas (Gasparini et al.,

1999).

In light of these considerations, very recently, Marangolo and colleagues (2018) aimed to verify whether the cerebellum, as part of the motor system, contributes to verb retrieval. In particular, the authors wanted to investigate if cerebellar stimulation would result efficacious for any given task or only if the task would require cognitive effort, or the involvement of other cognitive components, such as working memory, interacting with language. Indeed, several studies have already supported the hypothesis that the role of the cerebellum in language processing depends on task demands (Pope & Miall, 2014; Stoodley et al., 2010, 2012; Ackermann et al., 2007). Ackermann et al. (2007) have argued that non-linguistic aspects of task performance, such as the amount of effort or the degree of automaticity, might account for cerebellar involvement during verb generation. Similarly, Stoodley and Schmahmann (2009) have claimed that the cerebellum takes part not in the language function per sé but only when the task is cognitively demanding and, thus, engages other cognitive components, such as working memory and/or executive functions (Stoodley & Schmahmann, 2009). Indeed, apart from motor control and higher order aspects of speech production, a variety of studies have pointed to a contribution of the cerebellum in executive and memory tasks (Ackermann et al., 2007). Because the paradigm of verb generation involves the production and selection of different verbal responses (Thompson-Schill et al., 1998), pre-articulatory rehearsal processes are engaged as well, which rely to working memory processes (Ackermann et al., 2007; Helmuth, Ivry, & Shimizu, 1997).

Following this hypothesis, Marangolo et al (2018) investigated the effect of cerebellar DCS coupled with a verb training in 12 aphasic individuals by contrasting two different language tasks with different demands in terms of cognitive effort: a verb naming and a verb generation task. Indeed, with respect to verb naming in which the production of the correct answer is facilitated by the presented picture, verb generation, because of some combination of both retrieval and competition demands (Snyder, Banich, & Munakata, 2011), relies on different cognitive strategies (Justus, Ravizza, Fiez, & Ivry, 2005; Ackermann, Mathiak, & Riecker, 2007).

Like verb naming, verb generation is a semantic association task in which the participant has to produce a verb strictly associated to a given noun. Much of the cognitive demand between the two tasks is shared, including semantic and lexical retrieval processes and the planning, execution, and monitoring of speech production (e.g., Levelt, 1989). However, whereas in verb naming, the correct answer is univocally determined by the presented picture and the task is one of the earliest linguistic skills developmentally mastered and, thus, is an overlearned task (Herholz et al., 1997), verb generation requires the patient to creatively link a noun to a verb choosing among competing response alternatives (Thompson-Schill et al.,

1998). Interestingly, although verb generation is a task more cognitive demanding than verb naming and persons with aphasia generally experience greatest difficulty with verb generation (Martin & Cheng, 2006; Thompson-Schill et al., 1998), in Marangolo et al.'s study (2018), aphasic patients benefited only for this task after right cerebellar cathodal stimulation. Because the results point to potential therapeutic benefits of cerebellar stimulation only for complex language tasks, the authors believe that these findings have important implications for aphasia. Indeed, they address the possibility that the cerebellum might support cognitive functions which sustain language recovery (Marangolo et al., 2018).

Spinal cord stimulation for verb recovery

Following the hypothesis of an involvement of the motor system in processing some aspects of speech and, in particular, action verbs, very recently, Marangolo et al. (2017) investigated whether the spinal cord might also take part in verb retrieval. Indeed, given the strong reciprocal connections between the cortex and the spinal cord, the authors assumed that the stimulation of the spinal cord would influence activity into the sensorimotor cortex, through its ascending spinal pathways, which, in turn, would facilitate verb processing. Thus, Marangolo et al. (2017) explored the combined effect of transcutaneous spinal direct current stimulation (tsDCS) and language treatment for the recovery of verbs and nouns in 14 chronic aphasics. During each treatment, each subject received tsDCS (20 min, 2 mA) over the thoracic vertebrae (10th vertebra) in three different conditions: (1) anodal, (2) cathodal and (3) sham, while performing a verb and noun naming tasks. Each experimental condition was run in five consecutive daily sessions over 3 weeks with 6 days of intersession interval. The findings showed that anodal tsDCS differently affected the amount of improvement in noun and verb naming. Indeed, while noun and verb naming significantly improved in all patients for each condition at the end of training due to language treatment, anodal tsDCS boosted recovery only for verbs. There were no significant differences for the recovery of nouns in the three experimental conditions. This specificity argues against an effect simply due to enhanced cognitive arousal which should have influenced both verb and noun naming. The hypothesis was advanced that anodal tsDCS might have influenced activity along the ascending somatosensory pathways, ultimately eliciting neurophysiological changes into the sensorimotor areas which, in turn, supported the retrieval of verbs (Marangolo et al., 2017). Conversely, nouns did not benefit from stimulation, because they were not denoted by sensorimotor features. Although this finding seems surprising and it requires further investigation, it suggests that the spinal cord might take part in language processing, acting as a "bridge" for conveying tsDCS induced changes into brain networks involved in verb naming.

Conclusion

In the present work, we have reviewed a series of DCS studies on verb processing which starting from the modularistic approach, derived from the earliest neuropsychological works on aphasic patients, have confirmed the involvement of the frontal cortex in verb processing. Moreover, several studies were reported, which, motivated by the "embodied cognition" perspective, have suggested that some less traditional areas, connected to the frontal cortex, such as the sensorimotor cortex, take part in verb processing (Buccino et al., 2004; Gazzola et al., 2007; Rizzolatti et al., 2009; Casile et al., 2011). More recent works have further argued that the network supporting verb processing extends beyond the motor cortex to its interconnected structures. Indeed, some studies have already documented that other structures, such as the cerebellum and the spinal cord, might act as ancillary systems for verb recovery.

Although future studies will further elucidate our understanding on the role of these structures in language recovery, we believe that these results might have important clinical implications. Indeed, given the wide variability of cortical lesions among aphasic patients, it is not always easy to localize through non-invasive brain stimulation techniques, such as tDCS, the optimal stimulation cortical sites, unless we use additional very expensive methodologies, such as neuroimaging and/or modeling. This points to the urgency of considering other vicarious systems, functionally connected to the brain, that, when stimulated, contribute to the recovery of language. Thus, these results are promising since, for the first time, they suggest that the modulation of structures functionally connected to the brain, such as the cerebellum and the spinal cord, influences the language pathways leading to the recovery of verbs. Since verbs play a crucial role in sentence construction which is essential to enhance speech production in persons with aphasia, we believe that this finding is important for treatment outcomes.

References

Ackermann, H., Mathiak, K., & Riecker, A. (2007). The contribution of the cerebellum to speech production and speech perception: clinical and functional imaging data. *The Cerebellum*, 6, 202– 13.

Aggujaro, S., Crepaldi, D., Pistarini, C., Taricco, M., & Luzzatti, C. (2006). Neuro-anatomical correlates of impaired retrieval of verbs and nouns: interaction of grammatical class, imageability and actionality. *J. Neurolinguistics 19*, 175–194.

Alyahya, R. S. W., Halai, A. D., Conroy, P., & Lambon Ralph, M. A. (2018). Noun and verb processing in aphasia: Behavioural profiles and neural correlates. *NeuroImage: Clinical, 18*, 215-230.

Bernardis, P., & Gentilucci, M. (2006). Speech and gesture share the same communication system. *Neuropsychologia, 44,* 178-190.

Binkofski, F., & Buccino, G. (2006). The role of ventral premotor cortex in action execution and action understanding. *Journal of Physiology-Paris, 99,* 396-405.

Bonifazi, S., Tomaiuolo, F., Altoè, G., Ceravolo, M.G., Provinciali, L., & Marangolo, P. (2013). Action observation as a useful approach for enhancing recovery of verb production: new evidence from aphasia. *European Journal of Physical and Rehabilitation Medicine, 49,* 473-481.

Branscheidt, M., Hoppe, J., Zwitserlood, P., et al. (2017). tDCS over the motor cortex improves lexical retrieval of action words in post-stroke aphasia. *Journal of neurophysiology, 119,* 621-630.

Broca, P.P. (1861). Loss of Speech, Chronic Softening, and Partial Destruction of the Anterior Left Lobe of the Brain. *Bulletin de la Société Anthropologique, 2,* 235-238.

Buccino, G., Binkofski, F., Fink, G. R., Fadiga, L., Fogassi, L., Gallese, V., ... & Freund, H. J. (2001). Action observation activates premotor and parietal areas in a somatotopic manner: an fMRI study. *European Journal of Neuroscience, 13,* 400-404.

Buccino, G., Lui, F., Canessa, N., Patteri, I., Lagravinese, G., Benuzzi, F., ... & Rizzolatti, G. (2004). Neural circuits involved in the recognition of actions performed by nonconspecifics: An fMRI study. *Journal of Cognitive Neuroscience, 16,* 114-126.

Cappa, S. F., & Perani D. (2002). Nouns and verbs: neurological correlates of linguistic processing. *Riv Linguistica, 14,* 73-83.

Caramazza, A., & Hillis, A. (1991). Lexical organization of nouns and verbs in the brain. *Nature, 349,* 788-790.

Casile, V.C., & Ferrari P.F. (2011). The Mirror Neuron System: A Fresh View. *Neuroscientist, 17,* 524-538.

Crepaldi, D., Berlingeri, M., Cattinelli, I., Borghese, N.A., Luzzatti, C., & Paulesu, E. (2013). Clustering the lexicon in the brain: a meta-analysis of the neurofunctional evidence on noun and verb processing. *Frontiers in Human Neuroscience, 7,* 1-15.

Daniele, A., Giustolisi, L., Silveri, MC., Colosimo, C., & Gainotti G. (1994). Evidence for a possible neuroanatomical basis for lexical processing of nouns and verbs. *Neuropsychologia, 32,* 1325-1341.

De Smet, H.J., Baillieux, H., De Deyn, P.P., Mariën, P., & Paquier, P. (2007). The cerebellum and language: The story so far. *Folia Phoniatrica et Logopaedica, 59,* 165-170.

Fabbro, F., Moretti, R., & Bava, A. (2000). Language impairments in patients with cerebellar lesions. *Journal of Neurolinguistics, 13,* 173-188.

Fiori, V., Cipollari, S., Di Paola, M., Razzano, C., Caltagirone, C., & Marangolo, P. (2013). tDCS stimulation segregates words in the brain: evidence from aphasia. *Frontiers in Human Neuroscience, 7,* 269.

Fischer, M. H., & Zwaan, R. A. (2008). Embodied language: A review of the role of the motor system in language comprehension. *The Quarterly Journal of Experimental Psychology, 61,* 825-850.

Gasparini, M., Di Piero, V., Ciccarelli, O., Cacioppo, M. M., Pantano, P., & Lenzi, G. L. (1999). Linguistic impairment after right cerebellar stroke: a case report. *European Journal of Neurology, 6,* 353–6.

Gazzola, V., Rizzolatti, G., Wicker, B., & Keysers, C. (2007). The anthropomorphic brain: the mirror neuron system responds to human and robotic actions. *Neuroimage, 35,* 1674-1684.

Gentilucci, M., & Dalla Volta, R. (2008a). Spoken language and arm gestures are controlled by the same motor control system. *The Quarterly Journal of Experimental Psychology, 61,* 944-957.

Gentilucci, M., Dalla Volta, R., & Gianelli, C. (2008b). When the hands speak. *Journal of Physiology Paris, 102,* 21-30.

Justus, T., Ravizza, S. M., Fiez, J. A., & Ivry, R. B. (2005). Reduced phonological similarity effects in patients with damage to the cerebellum. *Brain and Language, 95,* 304–318.

Kemmerer, D., & Gonzalez Castillo, J. (2010a). The Two-Level Theory of verb meaning: an approach to integrating the semantics of action with the mirror neuron system. *Brain and Language, 112,* 54-76.

Kemmerer, D., & Eggleston, A. (2010b). Nouns and verbs in the brain: Implications of linguistic typology for cognitive neuroscience- *Lingua, 120,* 2686-2690.

Kemmerer, D. (2014). Word classes in the brain: implications of linguistic typology for cognitive neuroscience. *Cortex 58,* 27-51.

Helmuth, L. L., Ivry, R. B., & Shimizu, N. (1997). Preserved performance by cerebellar patients on tests of word generation, discrimination learning, and attention. *Learning & Memory, 3,* 456-474.

Herholz, K., Reulen, J. H., von Stockhausen, H. M., Thiel, A., Ilmberger, J., Kessler, J., Eisner, W., Yousry, T. A., & Heiss, W. D. (1997). Preoperative activation and intraoperative stimulation of language-related areas in patients with glioma. *Neurosurgery, 41,* 1253-1260.

Laiacona, M., & Caramazza A. (2004) The noun/verb dissociation in language production: varieties of causes. *Cognitive Neuropsychology, 21,*103-123.

Leggio, M. G., Silveri, M. C., Petrosini, L., & Molinari, M. (2000). Phonological grouping is specifically affected in cerebellar patients: a verbal fluency study. *Journal of Neurology, Neurosurgery & Psychiatry, 69,* 102-106.

Levelt, W.J.M. (1989). *Speaking: From Intention to Articulation.* Cambridge, MA: MIT Press.

Li, P., Jin, Z., & Tan, L. (2004). Neural representations of nouns and verbs in Chinese: an fMRI study. *NeuroImage, 21,* 1533-1541.

Lichtheim, L. (1885). On aphasia. *Brain, VII,* 433-484.

Liuzzi, G., Freundlieb, N., Ridder, V., Hoppe, J., Heise, K., Zimerman, M., & Hummel, F.C. (2010). The involvement of the left motor cortex in learning of a novel action word lexicon. *Current Biology, 20,* 1745-1751.

Luzzatti, C., Aggujaro, S., & Crepaldi, D. (2006). Verb–noun double dissociation in aphasia: theoretical and neuroanatomical foundations. *Cortex, 42,* 875-883.

Marangolo, P., Bonifazi, S., Tomaiuolo, F., Craighero, L., Coccia, M., Altoè, G., & Cantagallo, A. (2010). Improving language without words: first evidence from aphasia. *Neuropsychologia, 48,* 3824-3833.

Marangolo, P., Cipollari, S., Fiori, V., Razzano, C., & Caltagirone, C. (2012). Walking but not barking improves verb recovery: implications for action observation treatment in aphasia rehabilitation. *PloS one, 7:* e38610.

Marangolo, P., Fiori, V., Shofany, J., Gili, T., Caltagirone, C., Cucuzza, G., & Priori, A. (2017). Moving Beyond the Brain: Transcutaneous Spinal Direct Current Stimulation in Post-Stroke Aphasia. *Frontiers in Neurology. 8,* 400.

Marangolo, P. (2017). The potential effects of transcranial direct current stimulation (tDCS) on language functioning: Combining neuromodulation and behavioural intervention in aphasia. *Neuroscience Letters.* doi.org/10.1016/j.neulet.2017.12.057.

Marangolo, P., Fiori, V., Caltagirone, C., Pisano, F., Priori, A. (2018). Transcranial Cerebellar Direct Current Stimulation Enhances Verb Generation but not Verb Naming in Poststroke Aphasia. *Journal of Cognitive Neuroscience, 30,* 188-199.

Mariën, P., Saerens, J., Nanhoe, R., Moens, E., Nagels, G., Pickut, B. et al. (1996). Cerebellar induced aphasia: Case report of cerebellar induced prefrontal aphasic language phenomena supported by SPECT findings. *Journal of the Neurological Sciences, 144,* 34-43.

Mariën, P., Engelborghs, S., Pickut, B., De Deyn, P. P. (2000). Aphasia following cerebellar damage: Fact or fallacy? *Journal of Neurolinguistics, 13,* 145-171.

Mariën, P., Ackermann, H., Adamaszek, M., Barwood, C. H., Beaton, A., Desmond, J. E., & Ziegler, W. (2014). Consensus paper: Language and the cerebellum: an ongoing enigma. *Cerebellum, 13,* 386-410.

Martin, R. C., & Cheng, Y. (2006). Selection demands versus association strength in the verb generation task. *Psychonomic Bulletin & Review, 13,* 396-401.

Meinzer, M., Darkow, R., Lindenberg, R., & Flöel, A. (2016). Electrical stimulation of the motor cortex enhances treatment outcome in post-stroke aphasia. *Brain, 139,* 1152-1163.

Miceli, G., Silveri, M. C., Nocentini, U., & Caramazza, A. (1988). Patterns of dissociation in comprehension and production of nouns and verbs. *Aphasiology, 2,* 351-358.

Michel, G.F., Babik, I., Nelson, E.L., Campbell, J.M., & Marcinowski, E.C. (2013). How the development of handedness could contribute to the development of language. *Developmental Psychobiology, 55,* 608-620.

Nitsche, M.A., & Paulus, W. (2011). Transcranial direct current stimulation–update 2011. *Restor Neurol Neurosci, 29*, 463-492.

Parsons, L. M., Fox, P.T., Downs, J. H., Glass, T., Hirsch, T. B., Martin, C. C., et al. (1995). Use of implicit motor imagery for visual shape discrimination as revealed by PET. *Nature, 375*, 54-58.

Piras, F. & Marangolo, P. (2007). Noun-verb naming in aphasia: a voxel-based lesion-symptom mapping study. *NeuroReport, 18*, 1455-1458.

Pope, P. A., & Miall, R. C. (2014). Restoring cognitive functions using non-invasive brain stimulation techniques in patients with cerebellar disorders. *Frontiers in Psychiatry, 5*, 1-7.

Pulvermüller, F., Hauk, O., Nikulin, V. V., & Ilmoniemi, R. J. (2005). Functional links between motor and language systems. *European Journal of Neuroscience, 21*, 793-797.

Richter, S., Kaiser, O., Hein-Kropp, C., Dimitrova, A., Gizewski, E., Beck, A., ... & Timmann, D. (2004). Preserved verb generation in patients with cerebellar atrophy. *Neuropsychologia, 42*, 1235-1246.

Rizzolatti, G., Fabbri-Destro, M., & Cattaneo, L. (2009). Mirror neurons and their clinical relevance. *Nature Clinical Practice Neurology, 5*, 24-34.

Schmahmann, J. D., & Sherman, J. C. (1998). The cerebellar cognitive affective syndrome. *Brain: a journal of neurology, 121*, 561-579.

Shapiro, K. A., Moo, L. R., & Caramazza, A. (2006) Cortical signatures of noun and verb production. *Proceedings of National Academy of Sciences, 103*, 1644-1649.

Siegel, J. S., Shulman, G. L., & Corbetta, M. (2017). Measuring functional connectivity in stroke: Approaches and considerations. *Journal of Cerebral Blood Flow & Metabolism.*
https://doi.org/10.1177/0271678X17709198,0271678X1770919.

Siri, S., Tettamanti, M., Cappa, S., Rosa, P., Saccuman, C., Scifo, P., & Vigliocco, G. (2008). The neural substrate of naming events: effects of processing demands but not of grammatical class. *Cerebral Cortex 18*, 171-177.

Snyder, H. R., Banich, M. T., & Munakata, Y. (2011). Choosing Our Words: Retrieval and Selection Processes Recruit Shared Neural Substrates in Left Ventrolateral Prefrontal Cortex. *Journal of Cognitive Neuroscience, 23*, 3470-3482.

Stoodley, C. J., & Schmahmann, J. D. (2009). The cerebellum and language: Evidence from patients with cerebellar degeneration. *Brain and Language, 110*, 149-153.

Stoodley, C. J., Valera, E. M., & Schmahmann, J. D. (2010). An fMRI study of intraindividual functional topography in the human cerebellum. *Behavioural Neurology, 23*, 65-79.

Stoodley, C. J., Valera, E. M., & Schmahmann, J. D. (2012). Functional topography of the cerebellum for motor and cognitive tasks: an fMRI study. *Neuroimage, 16*, 1560-1570.

Thompson-Schill, S. L., Swick, D., Farah, M. J., D'Esposito, M., Kan, I. P., & Knight, R. T.

(1998). Verb generation in patients with focal frontal lesions: a neuropsychological test of neuroimaging findings. *Proceedings of the National Academy of Sciences, 95,* 15855–60.

Tomasino B., Tronchin G., Marin D., Maieron M., Fabbro F., Cubelli R., Skrap M., & Luzzatti C. (2018). Noun–verb naming dissociation in neurosurgical patients. *Aphasiology,* doi: 10.1080/02687038.2018.1542658.

Tranel, D., Adolphs, R., Damasio, H., & Damasio, A.R. (2001). A neural basis for the retrieval of words for actions. *Cognitive Neuropsychology, 18,* 655-674.

Ulm, L., Copland, D. & Meinzer, M. (2018). A new era of systems neuroscience in aphasia? *Aphasiology 32,* 742–764.

Wernicke, C. (1874). *Der Aphasische symptomencomplex.* Breslau: Cohn and Weigert.

Willems, R. M., & Hagoort, P. (2007). Neural evidence for the interplay between language, gesture, and action: A review. *Brain and language, 101,* 278-289.

Left neglect dyslexia, unilateral spatial neglect and the effects of sensory and transcranial stimulations

Giuseppe Vallar[1,2]*, Damiano Crivelli[1]

[1]Department of Psychology & Milan Center for Neuroscience – NeuroMi, University of Milano-Bicocca, Milan, Italy.

[2]IRCCS Istituto Auxologico Italiano, Neuropsychological Laboratory, Milan, Italy

*giuseppe.vallar@unimib.it

Abstract

Left neglect dyslexia (ND) is a component deficit of the syndrome of spatial unilateral neglect (USN) that is frequently associated with manifestations of USN outside the reading domain, conjuring up a double dissociation of defi*cits, and* *p*rima facie suggesting the independence of these putative discrete components of USN. However, while the number of right-brain-damaged patients with left USN without ND is quite sizeable across studies, right-brain-damaged patients with left ND without any other manifestation of left USN are quite rare, some with bilateral damage, one with a developmental disorder. This asymmetry supports the suggestion that ND is an intrinsic component of extra-personal USN, that shares with it most of the affected processes and representations of near extra-personal space. In line with this conclusion, the severity of left ND has been repeatedly found to be related with that of other manifestations of extra-personal USN, outside the reading domain. Furthermore, the sensory lateralized stimulations (caloric and galvanic vestibular, visual optokinetic, neck muscle vibration, head and trunk rotation, visuo-motor adaptation to optical prisms that displace the visual scene laterally), and transcranial magnetic stimulations of the brain, overall modulate left ND, as other manifestations of extra-personal USN. ND may be temporarily improved, worsened or left unchanged, according to the directionality of the stimulation. These effects appear to involve primarily egocentric/viewer/centred representations of written material, both lexical and non-lexical in nature (words, nonwords, text). Importantly, results from experimental, single session, paradigms, have been used to develop successful rehabilitation paradigms of ND.

Unilateral spatial neglect

Unilateral spatial neglect is a neuropsychological disorder, more frequent after lesions in the right cerebral hemisphere, and typically involving the left side of space, contralateral to the side of the lesion (contralesional). The defining feature of USN is that patients fail: i) to orient towards the contralesional sides of extra-personal and personal spaces, and ii) to explicitly report events occurring in those sides of space. USN may occur independent of sensory and motor peripheral disorders, such as hemianopia, hemianesthesia, and hemiplegia, which, in turn, may be present after unilateral brain damage, in the absence of USN. These double dissociations between USN and primary motor and sensory deficits indicate that the deficit cannot be traced back to peripheral sensorimotor impairments – that can actually be exacerbated by USN itself –, being instead due to higher-order disorders of spatial attentional and representational systems. USN is currently conceived as a multi-component deficit, whose different signs often manifest in association, but may also occur independent of one another. This suggests that the largely unitary phenomenal experience of space is in fact supported by multiple independent, though related, systems, that physiologically operate in concert, and, through the continuous integration of sensory and motor information, build up updated representations of the body-in-space, and of the space around us. USN is regarded as a disorder of spatial awareness, involving specific sectors of space, typically the contralesional side, since there is evidence that "neglected" (namely, not detected, as indexed by defective report) events undergo processing up to the sematic level. This "processing without awareness" may be revealed by paradigms that do not require the active exploration of space and explicit report of events, such as optional choice , and semantic priming .

USN may involve internally generated images of spatial scenes, such as specific known landscapes, retrieved from long-term memory, and imagined from a given spatial vantage point: the "Piazza del Duomo di Milano" experiment , and the "Piazza dei 500" study, in a patient with a left USN confined to mental visual imagery , and this may be taken as a further argument against interpretations in terms of peripheral impairments of lower-level sensorimotor systems. On the other hand, while USN may involve different sensory domains (visual, auditory, somatosensory-proprioceptive), and may be considered in this respect as a multimodal or multisensory disorder, its more striking manifestations have been observed in the visual domain, and USN is overall more severe in this sensory modality . Notwithstanding this predominance of the visual aspects of USN, a number of studies have reported selective, and modality-specific manifestations of USN, with dissociations between modality-specific deficits, such as: visual vs. tactile-proprioceptive USN ; visual vs. auditory USN .

Neglect dyslexia

One of the manifestations of USN concerns language, and, in the light of the preponderance and of the severity of the manifestations of USN in the visual modality, it is not surprising that the linguistic dimension involved is written language, and the USN-related disorder is *Neglect Dyslexia* (ND). ND has long been known. At the end of the XIX century, the Austrian neurologist Arnold Pick described a patient (case #2), with a left hemianopia, who showed left ND, omitting the first (left-sided) words of each line, and anosognosia (unawareness) for left hemiplegia, one components of the USN syndrome. Another patient showed "...a tendency to ignore objects exposed on the left side....While reading the patient preferred to hold the paper to the right or she would read on the right side of the paper." The patient, confronted with left-sided and right-sided stimuli presented simultaneously, "invariably looked at the right". These early observations show a close association of ND with other manifestations of USN. Other early studies report reading disorders associated with left USN after right-brain damage. In Brain's British patients the reading disorder consisted in omitting words and letters and reading from right to left. Paterson & Zangwill's patient #1 "commonly failed to attend to the left-hand page in turning the pages of a book and in reading lines of disconnected words commonly omitted the first word or two" (loc. cit. p. 339).

Patients with ND, when reading passages of prose, sentences, or individual words may commit errors in the contralesional side of the stimulus (i.e., the text line, the word or letter string). Errors include omissions (e.g., education ⬛ tion; famiglia [family] ⬛miglia [miles]), substitutions (e.g., wine ⬛ mine; albero [tree] ⬛pobero), and, less frequently, additions (luna [moon]⬛ moluna). In the neuropsychological taxonomies of the dyslexias acquired after brain damage, ND is classified as a "peripheral dyslexia", that affects processes regarded as not "central", and sparing therefore the lexical and semantic levels of analyses. Accordingly, patients with left ND may exhibit lexical effects in reading, and show evidence of semantic processing. The evidence from both single case reports and group studies indicates that ND, in the majority of reported patients, concerns the left side of the stimulus (in the case of ND, a letter string), encoded in i) viewer/ego-centered, and ii) stimulus-centered maps. The neural basis of these coordinate frames involves a main contribution from right hemispheric activity, with the hemispheric asymmetry that features visuo-spatial USN in general. A third, higher-level, of encoding of the letter string is in a graphemic/word-centred map, with a main contribution from the left hemisphere, at least in patients with a typical cerebral functional organization; patients with this pattern of ND are much less frequent than those with a viewer, or stimulus-centred, deficit. The impairment, found in left ND after right brain damage, matches the general distinction between space and object/stimulus-based manifestations of left USN. Figures 1 and 2 summarize the locus of functional impairment in ND, and the three levels of impaired representation.

Figure. 1 The dual-route cascaded (DRC) model of visual word recognition and reading aloud. The grey area covering the first two stages of processing of written material denotes the locus where ND deficits occur. The early stages of processing of written letter strings are further specified in Figure 2.

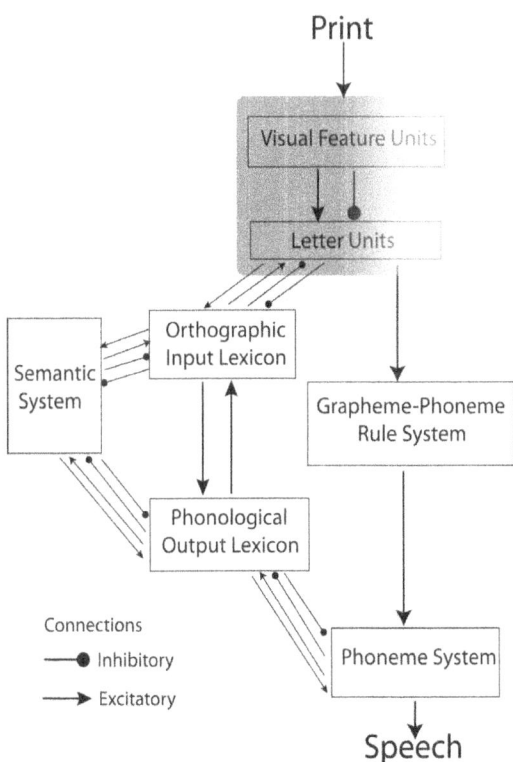

Figure 2. A model of ND including three levels of coordinate frames of a letter string, and two hemispheric patterns of the lateral distribution of spatial attention/representation. LH/RH: left/right hemisphere

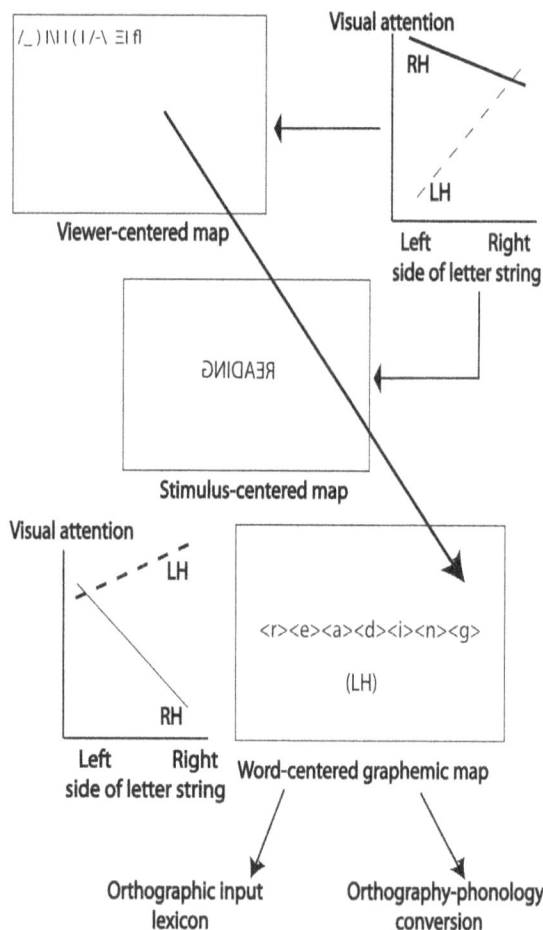

Unilateral spatial neglect and neglect dyslexia: Associations and dissociations

As shown in Table 1, right-brain-damaged patients with visuo-spatial USN, as assessed by target cancellation, line bisection, drawing, and perceptual judgement tasks, do not systematically show ND, which, in turn is infrequent without other associated manifestations of visuo-spatial USN. One of the studies summarized in Table 1 specifically aimed at assessing the frequency of ND, its association with other manifestations of contralesional USN, and the effects of visual field defects in a large sample of right-brain-damaged patients. The overall frequency of ND was about 20%, and over 35% in patients with USN, while ND without other manifestations of USN was very rare, amounting to one out of 138 patients.

The evidence from single case studies suggests that patients with left ND without other signs of USN may have lesion patterns divergent from the unilateral right hemispheric damage that characterizes most patients with left ND. The patient studied by De Lacy Costello & Warrington, who showed left ND and right USN, as assessed by drawing copy and line bisection tasks, had a large hyperdense mass in the parieto-occipital region of the left hemisphere, extending in the right hemisphere through the splenium of the corpus callosum. Another patient, with bilateral temporo-parietal vascular lesions, also involving the splenium of the corpus callosum, showed an error pattern characterized by substitutions of the initial (left) letters of words, nonwords and Arabic numbers, in stimulus-centred spatial coordinates, without USN in target cancellation, line bisection, and Landmark tasks, but left USN in the perceptual Wundt-Jastrow Illusion test. The stroke patient reported by Haywood & Coltheart had left ND without visuo-spatial USN; ND was however ipsilesional, since the lesion involved the left fronto-temporo-parietal region, but was possibly bilateral, affecting also the right hemisphere. Also the patients reported by Patterson & Wilson (1990), patient RR, with a deficit confined to the first letters of the words and by Katz & Sevush, presented with an ipsilesional ND reading deficit, associated with left brain damage. Finally, a child with no neurological deficit and focal brain damage showed a left developmental ND, involving the end of words, since Hebrew is read from right-to left, with no associated left visuo-spatial USN.

Taken together, findings from single case and group studies suggest both a close association between visuo-spatial USN, and ND, and the possibility of dissociations, compatible with the view that a spatial representational medium, with similar functional properties, may be articulated in discrete, although functionally related components . However, the two components of the double dissociation are largely asymmetrical. While patients with left visuo-spatial USN without left ND amount to at least 1/3 of the reported series [36% in the largest reported series of Lee et al.], patients with left ND without USN are most infrequent. Furthermore, in addition to the fact that the series of Lee et al. included only one ND patient with no other signs of USN in domains different from reading, the other reported patients in single case studies are neurologically or functionally non-typical, including bilateral lesions, as the patient reported by De Lacy Costello & Warrington, who showed neglect for the two opposite sides of space for written (left ND), and for visuo-spatial (right USN) material, and the patient reported by Arduino et al..

Since cases of ND without USN are very rare, and with behavioural and lesion patterns divergent from the canonical right hemispheric damage-left contralesional neglect deficit, these above discussed findings suggest a close relationship between ND and the manifestations of USN in non-linguistic domains. In an early report of six left-brain-damaged stroke and neoplastic patients with left ND, the reading deficit is most severe when also left USN is severe. Conversely, by inspecting their data from nine patients with left USN, Behrmann et al. do not detect any obvious association between the presence and the severity of USN and the reading profile. In larger series of right-brain-damaged patients (see Table I), significant positive correlations have been subsequently found between the severity of USN, as assessed by target cancellation, and line bisection tasks, and that of ND. In the group study with the largest sample size reported so far, the overall severity of visuo-spatial left USN (as assessed by target cancellation, bisection, and copy drawing tasks) is a significant predictor of the occurrence of left ND.

Table 1. Studies reporting right-brain-damaged stroke patients, assessed in a late subacute to chronic phase, showing and not showing (+/-) contralesional USN and ND, in reading words and nonwords.

Study	USN+ /ND+	USN+ /ND-	USN- / ND+	USN- /ND-	Mean duration of disease (months)	Stimuli
Ronchi N_{TOT}= 54	27	23	0	4	3,7	Word
Primativo N_{TOT}= 20	6	4	0	10	1,4	Word
Primativo N_{TOT}= 23	6	7	0	10	1,75	Word, Nonword

Arduino $N_{TOT}= 14$	8	6	0	0	2,5	Word, Nonword
Ptak $N_{TOT}= 54$	19 (+3)[#]	18	0	14	2,1	Word
Lee $N_{TOT}= 138$	30	54[$]	1	53	< 0,5	Word
Behrmann $N_{TOT}= 13$	5	4[%]	0	4[@]	11,6	Word

Note. [#]Three patients made only 1 ND reading error; [$]Four patients showed ipsilesional USN; [%]Four patients showed a left visual half-field deficit with USN and ND; [@]Four patients showed a left visual half-field deficit without USN or ND.

In line with the conclusion that ND is closely related to the other manifestations of extra-personal USN, a visual half-field deficit contralateral to the side of the hemispheric lesion (homonymous hemianopia) is not the basic deficit underlying ND. Since the early reports of left ND, it was apparent that left visual half-field deficits are frequently, but not systematically, associated with left ND. In the study by Kinsbourne & Warrington , two out of six right-brain-damaged patients with left ND showed no visual half-field deficits. All six patients with left ND reported by Arduino et al. had no visual field defects. Conversely, in the study by Behrmann et al. , all nine patients with left USN (five such patients also had left ND) had a left visual half-field deficit, while four hemianopic patients had no USN, including ND. In the large group study of Lee et al. 15 out of 30 (50%) patients with left USN and ND had a left visual field defect, that was present in four out of 50 (8%) USN patients without left ND. In the study by Ronchi et al. , 21 out of the 25 (84%) ND patients in whom the visual fields had been assessed showed a left visual field defect, that was present also in 11 out of 16 (69%) patients without left ND. Patients with left hemianopia without USN may not show ND .

Modulation of and by physiological and transcranial stimulations

In the light of the close association between left USN and left ND, the lateralized physiological stimulations that may temporarily ameliorate a variety of manifestations of USN, have been employed to attempt at improving also left ND. These studies have used either a single stimulation session approach, with predictable temporary effects, or rehabilitation paradigms. Results of these investigations are summarized in Table 2.

Table 2. Modulation of ND in right-brain-damaged patients by sensory and transcranial stimulations

STUDY	SAMPLE	STIMULATION & PROCEDURE	READING MATERIAL	EFFECT ON ND
Silberpfennig	1 (case #2) USN+, ND+	CVS (RE)	Word	▯ (RE)
Rubens	18 USN+	CVS (LE/RE, i/w)	Word	▯ (LEi/REw) ▯ (LEw/REi)
Rossetti	12 USN+ (Experiment 2)	PA-R (6 pts), NL (6 pts)	Text	▯ (PA-R), ▯ (NL)
Pizzamiglio	22 USN+	L-OKS+VST (11 Pts), VST (11 Pts), 6 weeks	Sentences (also copying)	▯ (no additive OKS effect)
Kerkhoff	10 USN+	L-OKS (5 Pts), VST (5 pts), 2 weeks	Paragraph	▯ (L-OKS)@ ▯ (VST)
Schindler	10 (5 USN+, 5 USN-)	20° L-HR, 20° L-TR, 20° R-HR, 20° R-TR	Paragraph	▯ (L-HR, L-TR) ▯ (R-HR, R-TR)
Farnè	6 USN+	PA-R, 1 session	Word, Nonword	▯ (PA-R)

Frassinetti	13 USN+	PA-R, 2 weeks (7 Pts) No-PA (6 Pts)	Word, Nonword	⊓ (PA-R)*
Schindler	20 USN+	(L-NMV+VST)-VST (10 pts) VST-(L-NMV+VST) (10 pts)	Paragraph	⊓(L-NMV+VST)
McIntosh	1 USN+	P A - R, 3 sessions	Poem	⊓
Angeli	13 USN+	PA-R (8 pts), NL (5 pts), 1 session	Word, Nonword	⊓ (PA-R), ⊓ (NL)
Rousseaux	10 USN+	PA-R, NL	Word, Nonword, Text	⊓ (PA-R, NL)
Schröder	30	L-OKS+VST (10 Pts), L-TENS+VST (10 Pts), VST (10 Pts), 4 weeks	Reading, writing a sentence	⊓(L-OKS+VST), ⊓(L-TENS+VST) ⊓ (VST)
Keller	10 USN+	L-OKS, L-OKS-PA, L-OKS-AM, VST, 1 session	Word	⊓ (L-OKS) ⊓ (L-OKS-AM, L-OKS-PA, VST)
Fortis	10 USN+	PA-R, 2 weeks	Sentence	⊓ (PA-R)
Reinhart	1 6 (9 U S N +, 7 USN-)	L-HR, R-HR	Paragraph	⊓ (L-HR) @ ⊓ (R-HR)
Reinhart	1 6 (9 U S N +, 7	L-OKS, R-OKS	Paragraph	⊓ (L-OKS) @

	USN-)			□ (R-OKS)
Cazzoli	24 USN+	cTBS-Sham (8 pts), Sham-cTBS (8 pts), No-Stim (8 Pts)	Text	□ (Sham-cTBS) □ (cTBS-Sham, No-Stim)
Oppenländer	24 (12 USN+ Text Copying)	GVS (CL/AR, AL/CR, Sham)	Text copying	□ (CL/AR) # □ (AL/CR, Sham)
Ptak	1 USN- (LB)$	PA-R	Word	□

Note.□/□/□: improved/worsened/unchanged. +/-: deficit present/absent (USN, ND); stimulation delivered/no stimulation delivered. L-/R-: left/right. Pts: patients; AM: right Arm Movements in the leftward direction; cTBS: continuous Theta Burst Stimulation; CVS: Caloric Vestibular Stimulation; i/w: iced/warm water; LE/RE: left/right ear; GVS: Galvanic Vestibular Stimulation; CL/AR: Cathodal Left/Anodal Right; AL/CR: Anodal Left/ Cathodal Right; HR/TR: Head/Trunk Rotation; LB: Line Bisection; NMV: Neck Muscle Vibration; OKS: visual OptoKinetic Stimulation; PA-R: Prism Adaptation to lenses displacing the visual scene rightward, with leftward AEs; NL: adaptation to Neutral Lenses; VST: Visual Scanning Training; @ Reduction of left-sided word omission errors; no effects on stimulus-based left-sided errors; * Significant effect on Nonword reading; Reduction of left-sided word omission errors; $ Significant leftward bias in LB.

Caloric Vestibular Stimulation (CVS)

An early observation concerns the effects of CVS on visuo-spatial USN and, presumably, ND: "While reading the patient preferred to hold the paper to the right or she would read on the right side of the paper" (loc. cit., p. 5). This patient made errors such as reading the middle letter of a word, as "M" in "name", or showing mirror reading ("Ave" for "Eva"). On caloric labyrinth stimulation of the right ear nystagmus to the left appeared, and, during the duration of it, the patient was able to read the word "Eva" correctly. After irrigation of the left ear, "Eva" was read "Ein". The effects of caloric stimulation on ND were unclear, both behaviourally, and with respect to the stimulated ear, and the effective caloric stimulation. Nevertheless, this was the first study suggesting a putative modulation of ND by vestibular stimulation.

Over 40 years later, Rubens assessed the effects of CVS on manifestations of USN in 18 right-brain-damaged patients, with a more definite and experimentally controlled approach. Before stimulation, patients failed to look consistently to the left to a verbal command, and to follow a finger moving from right to left over 10-15° over the midline, showed left USN in a line crossing test, and left ND in reading individual words. Positive effects of CVS on USN were found in 17 patients; one patient showed no caloric responses, and no improvement of USN. Thirteen out of the 17 patients who showed left ND in single word reading, being able to read only the last few right-sided letters of words, became able to read entire words. The remaining four patients improved their performance, but still omitted left-sided letters; these patients also showed the least improvement of the defective leftward gaze. The effective stimulations were ice water in the left external ear canal/warm water in the right canal, the latter being less effective. Patients were prevented to rotate their head in the direction of the slow phase of the nystagmus (leftward after the left ice/right warm ear stimulations, effective for USN and ND). Five minutes after CVS, the USN and ND deficit came back to the baseline, before stimulation, level. Patients made ND omission errors, for example, reading: [INVESTIG]ATE (before CVS), INVESTIGATE (immediately after CVS), [INVESTI]GATE" five minutes later]. Conversely, ice water in the right ear canal/left warm water in the left canal worsened the deficit in all 17 patients.

A somewhat related evidence comes from the observation by Oppenländer et al. , who found that subliminal galvanic vestibular stimulation (GVS) over the mastoids significantly improves both line bisection and text copying (left cathodal, right anodal GVS), and figure copying and digit cancellation (left anodal/right cathodal GVS). These findings on "neglect dysgraphia" are broadly in line with the abovementioned evidence of positive effects of CVS on ND.

Trunk and Head rotation (TR, HR)

The leftward rotation of the trunk transiently ameliorates manifestations of USN, putatively countering the ipsilesional pathological rightward rotation of the egocentric frame of reference, that may be a mechanism of USN. Schindler and Kerkhoff investigated the effect of TR and HR (20° rightward or leftward) in 10 right-brain damaged patients (five with USN without visual half-field deficits, five without USN). The reading task consisted in 15 irregularly indented paragraphs, including about 30 words, and 23-30 characters per line. ND errors (left-sided omissions and substitutions of words and syllables) were reduced by turning either the trunk or the head to the left, as compared to the baseline straight ahead condition, while a rightward rotation was ineffective. Similar effects were found on the rightward error in line bisection, that was reduced by trunk and head leftward rotation. In the five brain-damaged patients without USN no systematic effects on reading were found, and in five healthy control participants performance in the reading task was at ceiling. In a successive study, the differential effects of HR on viewer-based, egocentric (i.e., omissions of left-sided words), and word-based (omissions and substitutions of letters in the left-hand side of words) ND errors were assessed, using text paragraphs. Leftward HR reduces word omissions in the left hand-side of the text paragraph, but does not affect omissions and substitutions in the left hand-side of words, providing therefore evidence for effects on egocentric, spatial, rather than allocentric, stimulus-based, reference frames; rightward HR is ineffective.

Prism Adaptation (PA)

Both healthy and brain-damaged participants, when wearing optical prisms displacing the visual scene, while performing motor activities in extra-personal space, such as pointing to visual targets, adapt to the displacement, and, when the prisms are removed, show directional motor action biases towards the side opposite to the prism-induced displacement (aftereffects, AEs). PA is a procedure widely used to investigate behavioural and neural plasticity, in response to a distortion of the perceived visuo-spatial environment .

The first, pioneering, study showing positive effects of PA to optical prisms displacing the visual scene rightward, with leftward AEs, included the reading of a text. A single session of PA brought about a reduction of word omissions and alterations, as well as an improvement in other tasks assessing USN (line cancellation, line bisection, copy drawing), with effects persisting up to two hours . These findings were confirmed by a number of subsequent studies. The first study that used PA in a rehabilitation setting, with a two-week/ten sessions treatment, as compared to a control group of right brain-damaged USN patients not performing the PA treatment, was performed by Frassinetti et al. . In the PA group, the patients' performance improves in the Behavioral Inattention Test , that assesses, among other behavioural activities, reading (a menu), and activities with a reading component (telephone dialling, telling the time), and in reading nonwords . As for words, where errors are overall less -- a finding that may be taken as a further indication of preserved lexical processing in USN--, inspection of the means indicates an improvement too, that, however, is not significant. These findings have been confirmed by other rehabilitation studies from the same group, using the same PA paradigm , with different procedure of adaptation to the optical vision-displacing prims . No overall effects of a PA rehabilitation treatment on the BIT are also on record, in a single-blinded randomised controlled trial of PA, in which 28 right-brain-damaged stroke patients completed the trial, for improving self-care in stroke USN patients, although no specific mention of ND is made. In a study performed in 10 right-brain-damaged patients, Fortis et al. used an "ecological" procedure, with patients performing some daily activities (e.g., collecting coins), while wearing the prisms for PA in a two-week/ten session rehabilitation treatment of USN. The PA-induced improvement of USN was comparable to the one brought about by the repeated pointing activity, in a number of tasks, including sentence reading. However, while for cancellation performance a full mediational effect of prism exposure, as indexed by the AEs, was found on the improvement of patients', with larger AEs predicting a greater improvement in target cancellation tasks, this was not the case for sentence reading, suggesting differences between the effects of PA on these two manifestations of USN. All of these research studies however, while broadening knowledge regarding the effects on PA on various manifestations of USN, do not specifically focus on reading and dyslexia.

The relationship of the effects of PA on ND and other manifestations of USN was specifically investigated by Farnè et al. in six right-brain-damaged patients, assessing the effects of a single PA session. Both USN (target cancellation, line bisection) and ND (reading Italian 6-10 letter words, and nonwords, obtained substituting letters), improve, with comparable effects, that last up to 24 hours after PA, and fade after one week. In a subgroup of four patients a second PA session has comparable effects. A subsequent study based on these findings confirms that a single session of PA may improve ND, reducing left-sided omission errors in both word and nonword reading, with an overall effect of lexicality, performance level being higher with words. The amplitude of the first (left sided) saccade is also increased in the eight USN patients, who wear lenses displacing vision rightward, with leftward AEs.

Some negative findings are also on record, in studies focusing on USN and ND. In a single case study of a right-brain-damaged patient, three sessions of PA, spaced one week, improve USN in target cancellation, line bisection, and copying tasks, but not in a poem reading test, in which the patient's neglect errors are primarily left-sided omissions of whole words. Another more recent single case study concerns a right-brain-damaged patient, who, after having recovered from other aspects of visual USN, shows ND errors in reading individual words, while not making ND errors in text reading. The patient fails to detect leftward visual stimuli under conditions of single and double stimulation, and shows a leftward bias in line bisection, contrary to the standard behaviour of patients with left USN. In the experiment, individual words are presented for a limited (150 msec), rather than unlimited, as in previous studies showing positive effects of PA on left ND, amount of time. Under these conditions, baseline performance is very poor (10% correct), and only about 40% of reading errors can be traced back to left ND. PA brings about leftward AEs, but does not improve the reading disorder overall. The absence of other manifestations of left USN, and the error pattern during the experimental study, in which error rate is very high (90%), and prevailingly unrelated to neglect, may tentatively account for these negative findings. Specifically, the limited presentation time may have changed the features of the reading task, making it more of a detection task in a patient with a residual left visual field defect in the upper quadrant, than of an oculomotor exploratory task, that characterizes left ND, where leftward eye movements in reading are defective.

Finally, an entirely negative result from a group study is on record. In ten right-brain-damaged patients with left USN (in six left hemianopia is also present), a single session of PA fails to improve the deficit in target cancellation, line bisection, copy drawing, and in text reading, where ND errors are omissions of left-sided words.

Optokinetic Stimulation (OKS)

OKS involves the presentation of visual motion, such as a drum, or dots, which rotate, or move, in a specific direction (e.g., leftward, rightward) with respect to the viewer. OKS induces a physiological nystagmus . The OK nystagmus includes a linear or constant velocity slow phase, in the direction of the motion, and a quick return phase to the initial point of fixation just afterward.

OKS with a leftward direction of the movement temporarily diminishes the rightward bias in a number of tasks and improves a number of manifestations of the syndrome of USN. In patients with left USN, such effects, in the visuo-spatial domain, include the reduction of the disproportionate rightward deviation of the subjective visual straight ahead , and of the subjective centre of the segment in visual line bisection .

The rehabilitation study of Pizzamiglio et al. explored, in 22 right-brain-damaged patients with left USN, the effects of adding OKS to the behavioural treatment of USN, previously developed by their research group . This procedure makes use of a number of tasks (visual-spatial scanning, reading and copying training, copying of line drawings on a dot matrix, and figure description), that prompt the patient's active exploration of the neglected side of space. Patients were subdivided into two groups, one receiving the abovementioned scanning treatment , the other the treatment and an additional OKS. No further improvement of USN is found in the group of patients who receive the additional OKS. The behavioural assessment included sentence reading, that was not, however, specifically analysed.

In a successive study in 10 right brain-damaged patients with left USN, and nine with left ND, the effects of five sessions of leftward OKS were compared with those of a similar amount of visual scanning training, in two sub-groups of five patients. ND was assessed by a paragraph reading task, and two types of errors were measured: omissions of whole words and substitutions of words or of word parts, respectively considered as space- (with reference to egocentric, viewer-based, coordinate frames) and object/stimulus- (with reference to allocentric frames) based. While, contrary to previous evidence, the VST is not effective for rehabilitating USN, the OKS treatment improves the deficit, including reading. Whole word omissions decrease, while substitutions, both of the entire word, and of letters of it, considered as object-based errors, are unaffected by the OKS treatment; the improvement remains stable at a two-week follow up. These results were confirmed and extended in 30 right-brain-damaged patients with left USN, by the finding that both OKS (10 patients) and Transcutaneous Electrical Nerve Stimulation , delivered in combination with a VST for four weeks, improve both USN (as assessed by target cancellation, line bisection, figure copy and freehand drawing), ND, assessed by a reading test, and writing a sentence, the latter considered as a single factor. Conversely, the VST alone, also delivered for four weeks, is ineffective. The improvement in reading and writing, but not that on the standard tasks assessing USN, is still present at a one-week re-test. Additional support to the evidence that OKS improves left USN and ND (assessed by line bisection and target cancellation, and by word reading tasks), was provided in ten right-brain-damaged patients by a single-session study, assessing also the additional effects of PA and of movements of the right arm from right to left, compared to a visual scanning task as the control condition. The addition of PA is largely ineffective, as compared to the control condition, while that of right arm movement results in an increase, though not significant, of the rightward bias . In nine right-brain-damaged patients with left USN and ND, the effects of OKS on ND involve egocentric errors in paragraph reading, namely whole word omissions, that are reduced during OKS in a leftward direction, while rightward OKS is ineffective. Conversely, stimulus-centred errors (namely: omissions of left-sided letter(s), syllable(s) or half of a single word in compound words, such as "housekeeper" "keeper", and part-word substitutions, such as "mouse" "house") are not significantly affected by OKS. Broadly confirming evidence that OKS improves left ND, in two single case studies, leftward OKS was found to reduce omission, but not substitution, errors, in nonword reading.

Neck muscle vibration

The vibration of contralesional (left in right brain-damaged patients) posterior neck muscles (NMV) may temporarily improve left visual USN . In a rehabilitation study, left NMV was combined with a visual exploratory training of left USN. In a study using a cross-over design, the addition of NMV to 15 sessions of VST for three weeks brought about a reduction of a disproportionate rightward deviation of the subjective visual straight-ahead, and an improvement of left USN in visual (target cancellation), and in tactile exploratory tasks, as well as in an indented paragraph reading, respectively measured as targets detected and words read. These effects were larger than those associated with the visual exploratory training alone. The treatment gains in the different tasks assessing USN (target cancellation, exploration, text reading) are not correlated, suggesting that the treatments are effective on different components of USN . The correlations of these gains with the pre-treatment deviation of the straight-ahead during vibration are however significant, suggesting that the effects of the treatment primarily involve egocentric reference frames. This conclusion is supported also by the finding of a significant correlation of the variation of the straight-ahead between the end of the treatment and the eight weeks follow up (but not of the pre-treatment deviation of the straight-ahead during vibration) with those of the scores on the neglect tests, including reading, that nevertheless still do not correlate with each other.

Non-invasive brain stimulation (NIBS)

Non-invasive brain stimulations have been used to improve manifestations of USN, based on a model of hemispheric functioning after a unilateral lesion, that hypothesizes a disproportionate and maladaptive cerebral activity in the undamaged hemisphere, and a reduction of activity in the damaged hemisphere. The damaged hemisphere may be made further dysfunctional by the maladaptive hyperactivity of the undamaged one, via callosal interhemispheric inhibitory connections. According to this model, in right-brain-damaged patients with left USN, the left and the right hemispheres are, respectively, disproportionately active and exhibiting maladaptive activity, and hypoactive (in left-brain-damaged patients with aphasia, vice versa). Based on this model, a widely used approach involves the excitatory stimulation of the hemisphere damaged by the lesion responsible of the deficit of interest, the inhibitory stimulation of the undamaged, but dysfunctional and disproportionately active, hemisphere, or both. Overall, for the remediation of USN, NIBS are effective, particularly repetitive transcranial magnetic stimulation (rTMS), and, with less definite evidence, transcranial electrical direct current stimulation (tDCS), and continuous theta burst stimulation (cTBS). As for ND, in the study by Cazzoli et al. 24 stroke right brain-damaged patients underwent cTBS (eight trains of stimulation applied over two consecutive days on the contralesional, undamaged left posterior parietal cortex) in a randomized, double-blind, sham-controlled design, which also included a control group of neglect patients without stimulation. The study lasted three weeks, and, in the two groups of patients receiving cTBS, this was either preceded or followed by the Sham stimulation. cTBS exerts inhibitory effects that may outlast the stimulation period up to about 50 min . Reading was assessed by the Münich Reading Test (six 180-word parallel versions texts in German, with easy linguistic structure and short sentences). cTBS improves USN in activities of daily living, as assessed by the Catherine Bergego Scale , in the detection (also reducing response latencies) of left-sided visual targets, as assessed by the Vienna Test System, and in the two part Picture test, that requires the exploration of the drawing of two rooms, one left- and one right-sided, and the report of objects in it . Also reading improves, as indexed by the reduction of left-sided omissions, after cTBS preceded by Sham stimulation, but not vice versa (cTBS followed by Sham).

Concluding remarks

The experimental studies that have used the stimulations effective in ameliorating the variety of the non-linguistic manifestations of USN definitely show that ND is also improved, both with temporary effects, after single session stimulations, and in rehabilitation paradigms, with treatments typically lasting at least two weeks. This evidence, together with the very infrequent observation of ND without USN —while the opposite dissociation, namely USN without ND, is much more frequent— suggest that left ND, both in egocentric and in object/stimulus-based, allocentric reference frames (see Figure 2) should be considered as an intrinsic manifestation of extra-personal USN , that shares with unilateral extra-personal spatial deficits in the non-linguistic domain most of the attentional and representational processes involved. ND is then appropriately classified as a "peripheral" dyslexia, implying that the deficit may involve pre-lexical, and pre-semantic, spatial representations, in egocentric, or allocentric stimulus-based spatial reference frames, or in both of them (see Figure 2). In line with this conclusion, there is evidence that the stimulations that ameliorate extra-personal USN, also positively affect ND, although some negative findings using PA are on record. As for the frames of reference modulated by the stimulations, the evidence from different studies appears to indicate that egocentric, viewer-centred, representations (see Figure 1) are involved, with some studies, not showing effects on stimulus-based representations of lexical items.

References

Angeli, V., Benassi, M. G., & Làdavas, E. (2004). Recovery of oculo-motor bias in neglect patients after prism adaptation. *Neuropsychologia, 42*, 1223-1234.

Arduino, L. S., Burani, C., & Vallar, G. (2002). Lexical effects in left neglect dyslexia: A study in Italian patients. *Cognitive Neuropsychology, 19*, 421-444.

Arduino, L. S., Daini, R., & Silveri, M. C. (2005). A stimulus-centered reading disorder for words and numbers: Is it neglect dyslexia? *Neurocase, 11*, 405-415.

Arduino, L. S., Marinelli, C. V, Pasotti, F., Ferrè, E. R., & Bottini, G. (2012). Representational neglect for words as revealed by bisection tasks. *Journal of Neuropsychology, 6*, 43-64.

Azouvi, P., Olivier, S., de Montety, G., Samuel, C., Louis-Dreyfus, A., & Tesio, L. (2003). Behavioral assessment of unilateral neglect: study of the psychometric properties of the Catherine Bergego Scale. *Archives of Physical Medicine and Rehabilitation, 84,* 51-57.

Baron, R. M., & Kenny, D. A. (1986). The moderator-mediator variable distinction in social psychological research: conceptual, strategic, and statistical considerations. *Journal of Personality and Social Psychology, 51,* 1173-1182.

Behrmann, M., Black, S. E., McKeeff, T. J., & Barton, J. J. S. (2002). Oculographic analysis of word reading in hemispatial neglect. *Physiology & Behavior, 77,* 613-619.

Behrmann, M., & Moscovitch, M. (1994). Object-centered neglect in patients with unilateral neglect: Effects of left-right coordinates of objects. *Journal of Cognitive Neuroscience, 6,* 1-16.

Berti, A., Frassinetti, F., & Umiltà, C. A. (1994). Nonconscious reading? Evidence from neglect dyslexia. *Cortex, 30,* 181-197.

Berti, A., & Rizzolatti, G. (1992). Visual processing without awareness: Evidence from unilateral neglect. *Journal of Cognitive Neuroscience, 4,* 345-351.

Bisiach, E., Capitani, E., Luzzatti, C., & Perani, D. (1981). Brain and conscious representation of outside reality. *Neuropsychologia, 19,* 543-551.

Bisiach, E., Cornacchia, L., Sterzi, R., & Vallar, G. (1984). Disorders of perceived auditory lateralization after lesions of the right hemisphere. *Brain, 107,* 37-52.

Bisiach, E., & Luzzatti, C. (1978). Unilateral neglect of representational space. *Cortex, 14,* 129-133.

Bisiach, E., Meregalli, S., & Berti, A. (1990). Mechanisms of production control and belief fixation in human visuospatial processing: clinical evidence from hemispatial neglect and misrepresentation. In M. L. Commons, R. J. Herrnstein, S. M. Kosslyn, & D. B. Mumford (Eds.), *Quantitative analyses of behavior. Computational and clinical approaches to pattern recognition and concept formation:* Vol. IX (pp. 3-21). Hillsdale, New Jersey: Lawrence Erlbaum.

Brain, W. R. (1941). Visual disorientation with special reference to lesions of the right cerebral hemisphere. *Brain, 64,* 244-272.

Brandt, T., Dichgans, J., & Koenig, E. (1973). Differential effects of central verses peripheral vision on egocentric and exocentric motion perception. *Experimental Brain Research, 16,* 476-491.

Brunila, T., Jalas, M., Lindell, J. A., Tenovuo, O., & Hämäläinen, H. (2003). The two part picture in detection of visuospatial neglect. *The Clinical Neuropsychologist, 17,* 45-53.

Capitani, E., Neppi Modona, M., & Bisiach, E. (2000). Verbal-response and manual-response versions of the Milner Landmark task: normative data. *Cortex, 36,* 593-600.

Cazzoli, D., Müri, R. M., Schumacher, R., Von Arx, S., Chaves, S., Gutbrod, K., …

Nyffeler, T. (2012). Theta burst stimulation reduces disability during the activities of daily living in spatial neglect. *Brain, 135*, 3426-3439.

Coltheart, M., Rastle, K., Perry, C., Langdon, R., & Ziegler, J. (2001). DRC: a dual route cascaded model of visual word recognition and reading aloud. *Psychological Review, 108*, 204-256.

Cubelli, R., Nichelli, P., Bonito, V., De Tanti, A., & Inzaghi, M. G. (1991). Different patterns of dissociation in unilateral spatial neglect. *Brain and Cognition, 15*, 139-159.

Daini, R., Albonico, A., Malaspina, M., Martelli, M., Primativo, S., & Arduino, L. S. (2013). Dissociation in Optokinetic Stimulation Sensitivity between Omission and Substitution Reading Errors in Neglect Dyslexia. *Frontiers in Human Neuroscience, 7.*

De Lacy Costello, A., & Warrington, E. K. (1987). The dissociation of visuospatial neglect and neglect dyslexia. *Journal of Neurology, Neurosurgery, and Psychiatry, 50*, 1110-1116.

Denes, G., Crepaldi, D., & Zorzi, M. (2019). Dislessie e disgrafie acquisite. In G. Denes, L. Pizzamiglio, C. Guariglia, S. F. Cappa, D. Grossi, & C. Luzzatti (Eds.), *Manuale di neuropsicologia. Normalità e patologia dei processi cognitivi* (3rd ed., pp. 303-339). Bologna: Zanichelli.

Driver, J., & Pouget, A. (2000). Object-centered visual neglect, or relative egocentric neglect? *Journal of Cognitive Neuroscience, 12*, 542-545.

Fan, J., Li, Y., Yang, Y., Qu, Y., & Li, S. (2018). Efficacy of Noninvasive Brain Stimulation on Unilateral Neglect After Stroke. *American Journal of Physical Medicine & Rehabilitation, 97*, 261-269.

Farnè, A., Rossetti, Y., Toniolo, S., & Làdavas, E. (2002). Ameliorating neglect with prism adaptation: visuo-manual and visuo-verbal measures. *Neuropsychologia, 40*, 718-729.

Fortis, P., Maravita, A., Gallucci, M., Ronchi, R., Grassi, E., Senna, I., ... Vallar, G. (2010). Rehabilitating patients with left spatial neglect by prism exposure during a visuomotor activity. *Neuropsychology, 24*, 681-697.

Frassinetti, F., Angeli, V., Meneghello, F., Avanzi, S., & Làdavas, E. (2002). Long-lasting amelioration of visuospatial neglect by prism adaptation. *Brain, 125*, 608-623.

Friedmann, N., & Nachman-Katz, I. (2004). Developmental neglect dyslexia in a Hebrew-reading child. *Cortex, 40*, 301-313.

Gainotti, G. (2010). The role of automatic orienting of attention towards ipsilesional stimuli in non-visual (tactile and auditory) neglect: a critical review. *Cortex, 46*, 150-160.

Guariglia, C., Padovani, A., Pantano, P., & Pizzamiglio, L. (1993). Unilateral neglect restricted to visual imagery. *Nature, 364*, 235-237.

Gutschalk, A., & Dykstra, A. (2015). Auditory neglect and related disorders. In G. G.

Celesia & G. Hickok (Eds.), *Handbook of Clinical Neurology* (Vol. 129, pp. 557-571).

Halligan, P. W., Fink, G. R., Marshall, J. C., & Vallar, G. (2003). Spatial cognition: evidence from visual neglect. *Trends in Cognitive Sciences, 7,* 125-133.

Hamilton, R. H., Chrysikou, E. G., & Coslett, B. (2011). Mechanisms of aphasia recovery after stroke and the role of noninvasive brain stimulation. *Brain and Language, 118,* 40-50.

Haywood, M., & Coltheart, M. (2001). Neglect dyslexia with a stimulus-centered deficit and without visuospatial neglect. *Cognitive Neuropsychology, 18,* 577-615.

Hillis, A. E., & Caramazza, A. (1995). A framework for interpreting distinct patterns of hemispatial neglect. *Neurocase, 1,* 189-207.

Howard, I. P. (1982). *Human visual orientation.* Chichester: Wiley.

Jacobs, S., Brozzoli, C., Hadj-Bouziane, F., Meunier, M., & Farnè, A. (2011). Studying Multisensory Processing and Its Role in the Representation of Space through Pathological and Physiological Crossmodal Extinction. *Frontiers in Psychology, 2:* 89.

Jacquin-Courtois, S. (2015). Hemi-spatial neglect rehabilitation using non-invasive brain stimulation: or how to modulate the disconnection syndrome? *Annals of Physical and Rehabilitation Medicine, 58,* 251-258.

Jacquin-Courtois, S., O'Shea, J., Luauté, J., Pisella, L., Revol, P., Mizuno, K., ... Rossetti, Y. (2013). Rehabilitation of spatial neglect by prism adaptation: a peculiar expansion of sensorimotor after-effects to spatial cognition. *Neuroscience and Biobehavioral Reviews, 37,* 594-609.

Karnath, H.-O. (1996). Optokinetic stimulation influences the disturbed perception of body orientation in spatial neglect. *Journal of Neurology, Neurosurgery, and Psychiatry, 60,* 217-20.

Karnath, H.-O. (1997). Spatial orientation and the representation of space with parietal lobe lesions. *Philosophical Transactions of the Royal Society B: Biological Sciences, 352,* 1411-1419.

Karnath, H.-O., Christ, K., & Hartje, W. (1993). Decrease of contralateral neglect by neck muscle vibration and spatial orientation of trunk midline. *Brain, 116,* 383-396.

Karnath, H.-O., Fetter, M., & Dichgans, J. (1996). Ocular exploration of space as a function of neck proprioceptive and vestibular input -observations in normal subjects and patients with spatial neglect after parietal lesions. *Experimental Brain Research, 109,* 333-342.

Karnath, H.-O., Schenkel, P., & Fischer, B. (1991). Trunk orientation as the determining factor of the "contralateral deficit" in the neglect syndrome and as the physical anchor of the internal representation of body orientation in space. *Brain, 114,* 1997-2014.

Kashiwagi, F. T., El Dib, R., Gomaa, H., Gawish, N., Suzumura, E. A., da Silva, T. R., ... Bazan, R. (2018). Noninvasive brain stimulations for unilateral spatial neglect after stroke: A systematic review and meta-analysis of randomized and nonrandomized controlled trials. *Neural Plasticity*. https://doi.org/10.1155/2018/1638763

Katz, R. B., & Sevush, S. (1989). Positional dyslexia. *Brain and Language, 37*, 266-289.

Keller, I., Lefin-Rank, G., Lösch, J., & Kerkhoff, G. (2009). Combination of pursuit eye movement training with prism adaptation and arm movements in neglect therapy: A pilot study. *Neurorehabilitation and Neural Repair, 23*, 58-66.

Kerkhoff, G. (1998). Rehabilitation of visuospatial cognition and visual exploration in neglect: a cross-over study. *Restorative Neurology & Neuroscience, 12*, 27-40.

Kerkhoff, G., Keller, I., Ritter, V., & Marquardt, C. (2006). Repetitive optokinetic stimulation induces lasting recovery from visual neglect. *Restorative Neurology & Neuroscience, 24*, 357-369.

Kinsbourne, M., & Warrington, E. K. (1962). A variety of reading disability associated with right hemisphere lesions. *Journal of Neurology Neurosurgery and Psychiatry, 25*, 339-344.

Làdavas, E., Bonifazi, S., Catena, L., & Serino, A. (2011). Neglect rehabilitation by prism adaptation: different procedures have different impacts. *Neuropsychologia, 49*, 1136-1145.

Làdavas, E., Shallice, T., & Zanella, M. T. (1997). Preserved semantic access in neglect dyslexia. *Neuropsychologia, 35*, 257-270.

Lee, B. H., Suh, M. K., Kim, E. J., Seo, S. W., Choi, K.-M., Kim, G. M., ... Na, D. L. (2009). Neglect dyslexia: frequency, association with other hemispatial neglects, and lesion localization. *Neuropsychologia, 47*, 704-710.

Lowe, C. J., Manocchio, F., Safati, A. B., & Hall, P. A. (2018). The effects of theta burst stimulation (TBS) targeting the prefrontal cortex on executive functioning: A systematic review and meta-analysis. *Neuropsychologia, 111*, 344-359.

Marshall, J. C., & Halligan, P. W. (1988). Blindsight and insight in visuo-spatial neglect. *Nature, 336*, 766-767.

Massironi, M., Antonucci, G., Pizzamiglio, L., Vitale, M. V., & Zoccolotti, P. L. (1988). The Wundt-Jastrow illusion in the study of spatial hemi-inattention. *Neuropsychologia, 26*, 161-166.

Mattingley, J. B., Bradshaw, J. L., & Bradshaw, J. A. (1994). Horizontal visual motion modulates focal attention in left unilateral spatial neglect. *Journal of Neurology Neurosurgery and Psychiatry, 57*, 1228-1235.

McIntosh, R. D., Rossetti, Y., & Milner, A. D. (2002). Prism adaptation improves chronic visual and haptic neglect: a single case study. *Cortex, 38*, 309-320.

Oppenländer, K., Keller, I., Karbach, J., Schindler, I., & Reinhart, S. (2015). Subliminal galvanic-vestibular stimulation influences ego- and object-centred components of visual neglect. *Neuropsychologia, 74*, 170-177.

Ota, H., Fujii, T., Suzuki, K., Fukatsu, R., & Yamadori, A. (2001). Dissociation of body-centered and stimulus-centered representations in unilateral neglect. *Neurology, 57*, 2064-2069.

Ota, H., Fujii, T., Tabuchi, M., Sato, K., Saito, J., & Yamadori, A. (2003). Different spatial processing for stimulus-centered and body-centered representations. *Neurology, 60*, 1846-8.

Papagno, C., & Vallar, G. (2003). Anosognosia for left hemiplegia: Babinski's (1914) cases. In C. Code, C.-W. Wallesch, Y. Joanette, & A. R. Lecours (Eds.), *Classic Cases in Neuropsychology* (Vol. 2, pp. 171–189). Hove, East Sussex: Psychology Press.

Paterson, A., & Zangwill, O. L. (1944). Disorders of visual space perception associated with lesions of the right cerebral hemisphere. *Brain, 67*, 331-335.

Patterson, K., & Wilson, B. (1990). A ROSE is a ROSE or a NOSE: A deficit in initial letter identification. *Cognitive Neuropsychology, 7*, 447-477.

Pick, A. (1898). *Beiträge zur Pathologie und pathologischen Anatomie des Centralnervensystems, mit Bemerkungen zur normalen Anatomie desselben.* Berlin: Karger.

Pizzamiglio, L., Antonucci, G., Judica, A., Montenero, P., Razzano, C., & Zoccolotti, P. (1992). Cognitive rehabilitation of the hemineglect disorder in chronic patients with unilateral right brain damage. *Journal of Clinical and Experimental Neuropsychology, 14*, 901-923.

Pizzamiglio, L., Fasotti, L., Jehkonen, M., Antonucci, G., Magnotti, L., Boelen, D., & Asa, S. (2004). The use of optokinetic stimulation in rehabilitation of the hemineglect disorder. *Cortex, 40*, 441-50.

Pizzamiglio, L., Frasca, R., Guariglia, C., Incoccia, C., & Antonucci, G. (1990). Effect of optokinetic stimulation in patients with visual neglect. *Cortex, 26*, 535-540.

Primativo, S., Arduino, L. S., Daini, R., De Luca, M., Toneatto, C., & Martelli, M. (2015). Impaired oculo-motor behaviour affects both reading and scene perception in neglect patients. *Neuropsychologia, 70*, 90-106.

Primativo, S., Arduino, L. S., De Luca, M., Daini, R., & Martelli, M. (2013). Neglect dyslexia: a matter of "good looking." *Neuropsychologia, 51*, 2109-2119.

Ptak, R. (2017). What role for prism adaptation in the rehabilitation of pure neglect dyslexia? *Neurocase, 23*, 193-200.

Ptak, R., Di Pietro, M., & Schnider, A. (2012). The neural correlates of object-centered processing in reading: A lesion study of neglect dyslexia. *Neuropsychologia, 50*, 1142-1150.

Reinhart, S., Keller, I., & Kerkhoff, G. (2010). Effects of head rotation on space- and word-based reading errors in spatial neglect. *Neuropsychologia, 48*, 3706-

3714.

Reinhart, S., Schaadt, A. K., Adams, M., Leonhardt, E., & Kerkhoff, G. (2013). The frequency and significance of the word length effect in neglect dyslexia. *Neuropsychologia, 51,* 1273-1278.

Reinhart, S., Schindler, I., & Kerkhoff, G. (2011). Optokinetic stimulation affects word omissions but not stimulus-centered reading errors in paragraph reading in neglect dyslexia. *Neuropsychologia, 49,* 2728-2735.

Rizzolatti, G., Fadiga, L., Fogassi, L., & Gallese, V. (1997). The space around us. *Science, 277,* 190-191.

Ronchi, R., Algeri, L., Chiapella, L., Gallucci, M., Spada, M. S., & Vallar, G. (2016). Left neglect dyslexia: perseveration and reading error types. *Neuropsychologia, 89,* 453-464.

Rossetti, Y., Kitazawa, S., & Nijboer, T. (2019). Prism adaptation: From rehabilitation to neural bases. *Cortex, 111,* A1-A6.

Rossetti, Y., Rode, G., Pisella, L., Farné, A., Li, L., Boisson, D., & Perenin, M. T. (1998). Prism adaptation to a rightward optical deviation rehabilitates left hemispatial neglect. *Nature, 395,* 166-169.

Rossi, S., & Rossini, P. M. (2004). TMS in cognitive plasticity and the potential for rehabilitation. *Trends in Cognitive Sciences, 8,* 273-279.

Rousseaux, M., Bernati, T., Saj, A., & Kozlowski, O. (2006). Ineffectiveness of prism adaptation on spatial neglect signs. *Stroke, 37,* 542-543.

Rubens, A. B. (1985). Caloric stimulation and unilateral visual neglect. *Neurology, 35,* 1019-1024.

Schenkenberg, T., Bradford, D. C., & Ajax, E. T. (1980). Line bisection and unilateral visual neglect in patients with neurologic impairment. *Neurology, 30,* 509-517.

Schindler, I., & Kerkhoff, G. (1997). Head and trunk orientation modulate visual neglect. *Neuroreport, 8,* 2681-2685.

Schindler, I., Kerkhoff, G., Karnath, H.-O., Keller, I., & Goldenberg, G. (2002). Neck muscle vibration induces lasting recovery in spatial neglect. *Journal of Neurology Neurosurgery and Psychiatry, 73,* 412-419.

Schröder, A., Wist, E. R., & Hömberg, V. (2008). TENS and optokinetic stimulation in neglect therapy after cerebrovascular accident: a randomized controlled study. *European Journal of Neurology, 15,* 922-927.

Serino, A., Angeli, V., Frassinetti, F., & Làdavas, E. (2006). Mechanisms underlying neglect recovery after prism adaptation. *Neuropsychologia, 44,* 1068-1078.

Serino, A., Bonifazi, S., Pierfederici, L., & Làdavas, E. (2007). Neglect treatment by prism adaptation: what recovers and for how long. *Neuropsychological Rehabilitation, 17,* 657-687.

Siéroff, E. (2017). Acquired spatial dyslexia. *Annals of Physical and Rehabilitation Medicine, 60,* 155-159.

Silberpfennig, J. (1941). Contribution to the problem of eye movements. III. Disturbances of ocular movements with pseudo hemianopsia in frontal tumors. *Confinia Neurologica, 4*, 1-13.

Subbiah, I., & Caramazza, A. (2000). Stimulus-centered neglect in reading and object recognition. *Neurocase, 6*, 13-31.

Todman, D. (2009). Arnold Pick (1851-1924). *Journal of Neurology, 256*, 504-505.

Turton, A. J., O'Leary, K., Gabb, J., Woodward, R., & Gilchrist, I. D. (2010). A single blinded randomised controlled pilot trial of prism adaptation for improving self-care in stroke patients with neglect. *Neuropsychological Rehabilitation, 20*, 180-196.

Vallar, G. (1994). Left spatial hemineglect: An unmanageable explosion of dissociations? No. *Neuropsychological Rehabilitation, 4*, 209-212.

Vallar, G. (1998). Spatial hemineglect in humans. *Trends in Cognitive Sciences, 2*, 87-97.

Vallar, G. (1999). The methodological foundations of neuropsychology. In G. Denes & L. Pizzamiglio (Eds.), *Handbook of clinical and experimental neuropsychology* (pp. 95–131). Hove, East Sussex: Psychology Press.

Vallar, G. (2015). La stimolazione cerebrale non invasiva nella riabilitazione neuropsicologica. In N. Bolognini & G. Vallar (Eds.), *Stimolare il cervello. Manuale di stimolazione cerebrale non invasiva* (1st ed., pp. 119-138). Bologna: Il Mulino.

Vallar, G., & Bolognini, N. (2014). Unilateral spatial neglect. Oxford Handbooks Online. In A. C. (Kia) Nobre & S. Kastner (Eds.), *The Oxford Handbook of Attention* (pp. 1-77).

Vallar, G., Burani, C., & Arduino, L. S. (2010). Neglect dyslexia A review of the neuropsychological literature. *Experimental Brain Research, 206*, 219-235.

Vallar, G., & Calzolari, E. (2018). Unilateral spatial neglect after posterior parietal damage. In G. Vallar & H. B. Coslett (Eds.), *The Parietal Lobe. Handbook of Clinical Neurology* (Vol. 151, pp. 287-312).

Vallar, G., Guariglia, C., Magnotti, L., & Pizzamiglio, L. (1997). Dissociation between position sense and visual-spatial components of hemineglect through a specific rehabilitation treatment. *Journal of Clinical and Experimental Neuropsychology, 19*, 763-771.

Vallar, G., Guariglia, C., Nico, D., & Tabossi, P. (1996). Left neglect dyslexia and the processing of neglected information. *Journal of Clinical and Experimental Neuropsychology, 18*, 733-746.

Vallar, G., Guariglia, C., & Rusconi, M. L. (1997). Modulation of the neglect syndrome by sensory stimulation. In P. Thier & H.-O. Karnath (Eds.), *Parietal lobe contributions to orientation in 3D space* (pp. 555-578). Heidelberg: Springer Verlag.

Vallar, G., & Maravita, A. (2009). Personal and extra-personal spatial perception. In

G. G. Berntson & J. T. Cacioppo (Eds.), *Handbook of neuroscience for the behavioral sciences* (Vol. 1, pp. 322-336). New York: John Wiley & Sons.

Vallar, G., & Ronchi, R. (2019). Negligenza spaziale unilaterale e altri disordini unilaterali di rappresentazione. In G. Denes, L. Pizzamiglio, C. Guariglia, S. F. Cappa, & C. Luzzatti (Eds.), *Manuale di Neuropsicologia. Normalità e Patologia dei Processi Cognitivi* (3rd ed., pp. 577-626). Bologna: Zanichelli.

Vallar, G., Rusconi, M. L., Barozzi, S., Bernardini, B., Ovadia, D., Papagno, C., & Cesarani, A. (1995). Improvement of left visuo-spatial hemineglect by left-sided transcutaneous electrical stimulation. *Neuropsychologia, 33*, 73-82.

Vallar, G., Rusconi, M. L., Geminiani, G., Berti, A., & Cappa, S. F. (1991). Visual and nonvisual neglect after unilateral brain lesions: Modulation by visual input. *International Journal of Neuroscience, 61*, 229-239.

Wilson, B., Cockburn, J., & Halligan, P. W. (1987). *Behavioural inattention test.* Titchfield, Hants: Thames Valley Test Company.

Wischnewski, M., & Schutter, D. J. L. G. (2015). Efficacy and Time Course of Theta Burst Stimulation in Healthy Humans. *Brain Stimulation, 8*, 685-692.

Zihl, J. (2000). *Rehabilitation of visual disorders after brain injury.* Hove, United Kingdom: Psychology Press.

The importance of being earnest in earnestness: The influence of root morphology in complex noun processing

Maximiliano A. Wilson[1]*, Claudia Sánchez-Gutiérrez[2], Hugo Mailhot[3], S. Hélène Deacon[4]

[1] Centre de recherche CERVO et Département de réadaptation, Université Laval, Québec, Canada.

[2] Department of Spanish and Portuguese, University of California, Davis, CA, United States.

[3] Department of Computer Science, University of California, Davis, CA, United States.

[4] Dalhousie University, Halifax, Nova Scotia, Canada

*maximiliano.wilson@fmed.ulaval.ca

Abstract
The ability to identify the morphemes that compose a word facilitates its recognition. This is particularly relevant because most of the new words an adult reader will encounter are morphologically complex. The effects of root morphology (e.g., frequency and family size) have been widely investigated, with far less attention to prefix and suffix morphology. In the present study we explored the influence of prefix, root and suffix morphological variables (e.g., frequency, family size, productivity, etc.) on lexical decision (LD) latencies. We did so specifically for words that contain both a prefix and suffix. We used a regression model with RTs to lexical decisions to 1,228 morphologically complex English nouns that included both a suffix and a prefix. After controlling for the effect of lexical (Step 1) and semantic variables (Step 2), morphological variables affected latencies. Root cumulative frequency and prefix productivity exerted a facilitatory effect. The percentage of more frequent words than the target in the families of the prefix and the suffix had an inhibitory effect. Our results support the contribution of root frequency and, critically, they also extend previous findings on the influence of suffix morphology to prefix morphological variables. Our results shed new light on the importance of the lexical competitors of the families of prefixes and suffixes, beyond root morphological variables.

Morphemes play an important role in visual word recognition (see, Amenta & Crepaldi, 2012, for a review). Unlike root morphological effects, prefix and suffix morphology have been scarcely studied (Sánchez-Gutiérrez, Mailhot, Deacon, & Wilson, 2018). The aim of this study is to investigate the influence of prefix, root and suffix morphology in a regression study for lexical decision (LD) of morphologically complex English nouns that include both a prefix and a suffix (e.g., *un-conscious-ness*).

Morpheme frequency effects are interpreted as evidence of morphemic decomposition during word processing (Lehtonen, Niska, Wande, Niemi, & Laine, 2006). Root frequency effects have been consistently reported for LD (Burani, Salmaso, & Caramazza, 1984; Colé, Beauvillain, & Segui, 1989; Taft, 1979). Also, evidence of faster access to words that include roots from larger families (i.e. family size effects) than those from smaller families has also emerged (Balling & Baayen, 2008; Ford, Davis, & Marslen-Wilson, 2010; Moscoso del Prado Martín, Bertram, Häikiö, Schreuder, & Baayen, 2004). Ford et al. (2010) provided evidence of the independent effects of root frequency and morphological family size in word processing.

Unlike root morphology, the effects of prefix and suffix morphology have been scarcely studied. Burani and Thornton (2003) found that pseudowords that included frequent suffixes were more difficult to reject than pseudowords with low-frequency suffixes. In an LD study with words containing one suffix, Sánchez-Gutiérrez et al. (2018) found a facilitatory effect of several suffix variables, including frequency, length, family size and productivity. Conversely, the percentage of more frequent words in the family of the suffix had an inhibitory effect on latencies. Baayen, Wurm, & Ayckock (2007) have directly addressed prefix and suffix productivity in two regression studies. They found a facilitatory effect of affix productivity (Experiment 1) and affix family size (Experiment 2). However, they analyzed both prefix and suffix variables together. This prevents the study of the unique contribution of prefix and suffix morphological variables.

Thus, the individual contribution of prefix and suffix morphology to complex word processing remains to be studied. Here we addressed this issue by means of a LD study of a large set of English nouns with both a prefix and a suffix.

Method

Materials
We selected nouns out of the 3,622 words of the morphological database MorphoLex (Sánchez-Gutiérrez et al., 2018) that had one prefix, one root and one suffix. This resulted in 1,753 nouns. From these, we selected the words with information on LD latencies (RTs variable) in the English Lexicon Project (ELP; Balota et al., 2007). This left the list with 1,426 nouns. Afterwards, we checked the

availability of values for the psycholinguistics variables capturing the effects of critical control variables, specifically for lexical and semantic variables (see Steps 1 and 2). This resulted in a final list of 1,228 nouns with a prefix and a suffix. Table 1 shows the summary statistics for all the variables used in the LD study.

Data analysis

We inspected the data for skewness. Four variables showed skewness values greater than ±2 (Gravetter & Wallnau, 2014): frequency, orthographic neighborhood, root frequency and root family size. We log-transformed these variables, which resolved the skew problems (see Table 1).

We also inspected the data for multicollinearity. We calculated the tolerance and variance inflation factors (VIF). We found values indicating multicolinearity (VIF > 4 and tolerance < 0.2) for the following variables: frequency, prefix family size, the percentage of more frequent words in the morphological family of the prefix (prefix PFMF), prefix productivity (P*), root PFMF, suffix frequency, suffix family size, suffix PFMF and suffix P*. We log-transformed these variables (except for frequency, that was already transformed). This data transformation did not solve multicolinearity for several variables (see Table 1). To address this issue, we followed and ran additional regression models excluding one of these variables at a time.

Table 1. Summary statistics for all the variables used in the lexical decision study

	Mean	Standard Deviation	Minimum	Maximum	Skewness	Tolérance	VIF
Length	9.98	1.48	6	15	-0.006	0.54	1.84
log Frequency	2.73	0.89	0	5.38	-0.24	0.13	7.56'
log Orthographic neighborhood	0.09	0.16	0	0.69	1.47	0.77	1.28
Semantic neighborhood	665.13	665.97	1	3561	1.48	0.33	3.00
Number of meanings	2.60	1.79	0	12	1.78	0.67	1.49
log Prefix Frequency	5.61	.76	2.09	6.38	-1.42	0.47	2.10
log Prefix Family size	2.24	0.52	0.30	2.90	-1.18	0.21	4.78'
Prefix length	2.42	0.91	1	6	1.35	0.57	1.74
log Prefix PFMF	1.40	0.42	0	2.01	-1.12	0.23	4.34'
log Prefix P*	0.001	0.0007	0	0.002	-0.02	0.26	3.89
log Root Frequency	4.76	0.76	1.79	6.25	-0.85	0.34	2.90
log Root Family size	1.25	0.31	0.48	2.33	-0.03	0.43	2.31
log Root PFMF	1.49	0.44	0	2.004	-1.59	0.34	2.92
log Suffix Frequency	6.28	0.67	2.58	6.81	-1.99	0.07	13.55
log Suffix Family size	2.81	0.57	0.30	3.39	-1.43	0.06	15.93
Suffix length	2.85	0.77	1	5	-0.43	0.47	2.14
log Suffix PFMF	1.43	0.39	0	2.004	-0.98	0.11	9.29'
log Suffix P*	0.003	0.003	0	0.01	1.51	0.26	3.89
RTs	830.75	120.58	602	1320.33	0.82	n/a	n/a

Note. *Length*: whole word length in letters; *log Frequency*: log-transformed Hyperspace Analogue to Language (HAL) frequency; *log Orthographic neighborhood*: log-transformed orthographic neighborhood; *Semantic neighborhood*: log local semantic neighbourhood size; *Number of meanings*: log number of meanings; *log Prefix Frequency*: cumulative HAL frequency of the prefix; *log Prefix Family size:* log-transformed family size of the prefix; *Prefix length:* length in letters of the prefix; *log Prefix PFMF*: log-transformed percentage of more frequent words in the morphological family of the prefix; *log Prefix P**: log-transformed prefix productivity; *log Root Frequency:* log-transformed cumulative HAL frequency of the root; *log Root Family size*: log transformed family size of the root; *log Root PFMF:* log-transformed percentage of more frequent words in the morphological family of the root; *log Suffix Frequency*: cumulative HAL frequency of the suffix; *log Suffix Family size*: log-transformed family size of the suffix; *Suffix length*: length in letters of the suffix; *log Suffix PFMF*: log-transformed percentage of more frequent words in the morphological family of the suffix; *log Suffix P**: log-transformed suffix productivity; *RTs:* reaction times in milliseconds. Tolerance and variance inflation factor (VIF) are multicollinearity measures.

* Variables with Tolerance < .2 or VIF > 4 that indicate mulitcolinearity. These variables were excluded one by one in different regression models.

Following previous similar literature (Boukadi, Zouaidi, & Wilson, 2016; Cortese & Schock, 2013; Sánchez-Gutiérrez et al., 2018; Yap & Balota, 2009), we grouped and entered the variables in the regression models in three different steps. Step 1 included three lexical variables: length in letters, HAL (i.e., Hyperspace Analogue to Language) frequency and orthographic neighborhood size. We obtained the values of these variables from the ELP online database (elexicon.wustl.edu; Balota et al., 2007). Step 2 included two semantic variables. Semantic neighborhood (i.e., local semantic neighbourhood size) is a measure of the density of the semantic neighbourhood (Yap & Balota, 2009). The values for this variable were obtained from the WordMine2 database (uwindsor.ca/wordmine; Durda & Buchanan, 2006). The variable number of meanings refers to the number of meanings of a target word in its different synsets (i.e. sets of semantically related words; Baayen, Feldman, & Schreuder, 2006; Sánchez-Gutiérrez et al., 2018; Yap & Balota, 2009). Values for this variable were taken from the WordNet online database (wordnet.princeton.edu; Fellbaum, 1998). Both semantic variables were log-transformed. Step 3 included the morphological variables (n = 13): prefix frequency, prefix family size, prefix length, prefix PFMF, prefix P*, root frequency, root family size, root PFMF, suffix frequency, suffix family size, suffix length, suffix PFMF and suffix P*. Frequency refers to the summed token frequency of all members in the morphological family of a morpheme. Morphological family size is the number of word types in which a given morpheme is a constituent (Baayen et al., 2006). PFMF refers to the number of more frequent words in the family of a morpheme as compared to the target word. A value of 0 means that no word in the family is more frequent and 100 means that all words in the family are more frequent. For affix P*, values closer to 1 indicate high morphemic productivity, whereas values close to 0 indicate low productivity (Baayen & Renouf, 1996; Sánchez-Gutiérrez et al., 2018). The values for the morphological variables were obtained from the online database MorphoLex (https://github.com/hugomailhot/MorphoLex-en; Sánchez-Gutiérrez et al., 2018).

Results

Table 2 shows the results of the regression models. After controlling for the effect of lexical (Step 1) and semantic variables (Step 2), morphological variables (Step 3) explained a significant proportion of latencies. However, only root frequency was a significant predictor of LD RTs.

Table 2. Standardized βs, R²s, and ΔR²s for the regression analyses of lexical decision

Step	All variables	log Frequency	log Prefix PFMF	log Prefix Family size	log Suffix PFMF	log Suffix Family size	log Suffix Frequency
Step 1 Lexical variables							
Length	0.244***	.271***					
log Frequency	-.564***	n/a					
log Orthographic neighborhood	-.082***	-.122***					
R²	.422	.107					
ΔR²	.422***	.107***					
Step 2 Semantic variables							
Semantic neighborhood	-.144***	-.439***					
Number of meanings	-.067**	-.114***					
R²	.437	.364					
ΔR²	.015***	.257***					
Step 3 Morphological variables							
log Prefix Frequency	.016	-.023	.016	.008	.018	.017	.017
log Prefix Family size	-.030	-.019	-.030	n/a	-.021	-.028	-.028
Prefix length	-.023	-.032	-.023	-.022	-.026	-.023	-.023
log Prefix PFMF	.001	.096*	n/a	-.005	-.023	-.004	-.003
log Prefix P*	-.044	-.025	-.044	-.065*	-.051	-.046	-.046
log Root Frequency	-.132***	-.171***	-.132***	-.133***	-.122***	-.130***	-.130**
log Root Family size	.050	.051	.050	.051	.045	.050	.050
log Root PFMF	-.002	.025	-.002	-.002	-.019	-.000	-.002
log Suffix Frequency	.050	-.126	.050	.045	-.002	-.005	n/a
log Suffix Family size	-.065	.036	-.065	-.061	-.031	n/a	-.016
Suffix length	.052	.051	.052	.052	.052	.052	.054
log Suffix PFMF	-.109	.165**	-.109	-.104	n/a	-.097	-.092
log Suffix P*	-.048	.016	-.048	-.048	-.027	-.062	-.052
R²	.458	.429	.458	.457	.456	.457	.457
ΔR²	.020***	.066***	.020***	.020***	.019***	.020***	.020**

Note. *Length*: whole word length in letters; *log Frequency*: log-transformed Hyperspace Analogue to Language (HAL) frequency; *log Orthographic neighborhood:* log-transformed orthographic neighborhood; *Semantic neighborhood*: log local semantic neighbourhood size; *Number of meanings*: log number of meanings; *Prefix Frequency:* cumulative HAL frequency of the prefix; *log Prefix Family size*: log-transformed family size of the prefix; *Prefix length*: length in letters of the prefix; *log Prefix PFMF*: log-transformed percentage of more frequent words in the morphological family of the prefix; *log Prefix P**: log-transformed prefix productivity; *log Root Frequency*: log-transformed cumulative HAL frequency of the root; *log Root Family size*: log transformed family size of the root; *log Root PFMF:* log-transformed percentage of more frequent words in the morphological family of the root; *log Suffix Frequency:* cumulative HAL frequency of the suffix; *log Suffix Family size*: log-transformed family size of the suffix; *Suffix length*: length in letters of the suffix; *log Suffix PFMF:* log-transformed percentage of more frequent words in the morphological family of the suffix; *log Suffix P*:* log-transformed suffix productivity.

Columns refer to the different regression models. The first column refers to the full model that included all the variables. Subsequent columns refer to the variables that were excluded in the model due to collinearity.

ΔR^2 is the incremental increase in the model R2 that results from the addition of a predictor or set of predictors in a new step of the model.

*p < .05 ** p < .01 *** p < .001

It exerted a facilitatory effect on latencies: the higher the cumulative frequency of the morphological family of the root, the shorter the RTs. We then conducted a next set of analyses to address multicolinearity, excluding one variable that showed multicollinearity at a time. When frequency was excluded, prefix and suffix PFMF had an inhibitory effect on RTs. The larger the PFMF of the prefix and suffix, the longer the latencies. Also, when prefix family size was excluded from the model, prefix P* facilitated RTs. The higher the productivity of the prefix, the shorter the latencies.

Discussion

The aim of this study was to investigate morphological processing by means of a LD regression study with 1,228 English nouns with both a prefix and a suffix. We found that root cumulative frequency exerted a facilitatory effect on latencies. When we excluded other variables that showed multicolinearity from analyses, the percentage of more frequent words than the target in the families of the prefix and the suffix showed an inhibitory effect on latencies. In addition, prefix productivity exerted a facilitatory effect on latencies.

Our results indicate that root cumulative frequency has a facilitatory effect. This means that the higher the cumulative frequency of a root, the shorter the LD latencies, replicating prior studies (Burani & Thornton, 2003; Luke & Christianson, 2011; Sánchez-Gutiérrez et al., 2018; Taft, 1979). Baayen et al., (2007, Study 2) found that the effect of root frequency varied according to frequency: facilitating RTs for low frequency words and inhibiting RTs for higher frequency words. It is possible that our stimuli did not capture this whole range of word frequency. Even if the whole-word frequency of the stimuli of the present study is higher than the one in Baayen et al.'s study (mean HAL frequencies = 3,161 and 228, respectively, t (1,420) = 3.54, $p < .001$), words were not in the high-end of frequency either. In line with Baayen et al. and Sánchez-Gutiérrez et al., we found a facilitatory effect of both whole-word and root frequency. This provides further evidence of the central role of root frequency and extends its effects to words that contain both a prefix and a suffix.

We also found effects of some morphological features of the prefixes and suffixes. Specifically, the percentage of more frequent words than the target word in the morphological families of both the prefix and the suffix had an inhibitory effect. Thus, the higher the percentage of more frequent words than the target sharing the same prefix and suffix, the longer the latencies. These results are in line with our previous study with suffixed words, in which we found an inhibitory effect of this variable (Sánchez-Gutiérrez et al., 2018). We extend now the same results to prefixes. Taken together, these results suggest that a word with a considerable number of competitors (i.e. more frequent words) in its prefix and suffix morphological families is processed more slowly than a word that is amongst the most frequent ones in its prefix and suffix families.

Finally, we found a facilitatory effect of prefix productivity on latencies. The higher the productivity of the prefix, the shorter the latencies. Baayen et al. (2007) showed that affix (both prefix and suffix collapsed together) productivity facilitated latencies. In Sánchez-Gutiérrez et al. (2018) we found that suffix productivity facilitated word recognition. Here we extend these results to prefix productivity.

In the present study we failed to reproduce previous findings on the importance of suffix morphological variables for word recognition (Sánchez-Gutiérrez et al., 2018). Contrarily, we found here that root frequency was the strongest and almost sole predictor of LD latencies. The variables for both the present and previous studies were taken from the morphological English database MorphoLex (Sánchez-Gutiérrez et al., 2018). Thus, we can rule out different variable calculations as an explanation for these between-studies differences. Instead, one possible explanation may come from differences in whole-word frequency and in the salience of the roots of both studies. These results may be due to idiosyncratic characteristics of our items (Baayen et al., 2007). Morphologically complex words that contain both a prefix and a suffix, such as *prospectus,* tend to be less frequent words as compared with only suffixed words and are typically present in more formal language contexts. A comparison between both studies shows that the words used in Sánchez-Gutiérrez et al. (mean frequency = 4,814) are significantly more frequent than those used in the present study (mean = 3,161; t (5,950) = 3.43, $p < .001$). Additionally, the cumulative frequency of the roots in the present study (mean = 155,790) is also significantly higher than that of Sánchez-Gutiérrez et al. (mean = 61,837; t (5,950) = -17.07, $p < .001$). This means that the words in the present study were less frequent but contained roots that were more frequent or salient than the words in Sánchez-Gutiérrez et al. Baayen et al. (2007) state that, in the context of a highly-informative (or highly-frequent, in our case) roots, prefixes or suffixes become less relevant for word processing. This seems to be the case in the present study and might explain why we found that root cumulative frequency was the strongest predictor of LD latencies in the presence of both prefix and suffix morphological variables.

Our novel contribution here is the investigation of the simultaneous contribution of prefix, root and suffix morphological variables. We believe that this study represents a first step in addressing the processing of words with different morphological patterns, thus increasing the representativeness of findings to more words of the English language. Further studies need to pursue this line of research to understand morphological processing in its entirety.

Authors note: This research was supported by an Insight Development Grant awarded to M.A.W., S.H.D and C.S.G by the Social Sciences and Humanities Research Council (CRSH) of Canada, grant number: 430-2015-00699.

References

Amenta, S., & Crepaldi, D. (2012). Morphological processing as we know it: An analytical review of morphological effects in visual word identification. *Frontiers in Psychology, 3:* 1-12.

Baayen, R. H., Feldman, L. B., & Schreuder, R. (2006). Morphological influences on the recognition of monosyllabic monomorphemic words. *Journal of Memory and Language, 55,* 290-313.

Baayen, R. H., & Renouf, A. (1996). Chronicling the times: Productive lexical innovations in an English newspaper. *Language, 72,* 69-96.

Baayen, R. H., Wurm, L. H., & Aycock, J. (2007). Lexical dynamics for low-frequency complex words. A regression study across tasks and modalities. *The Mental Lexicon, 2,* 419-463.

Balling, L., & Baayen, R. H. (2008). Morphological effects in auditory word recognition: Evidence from Danish. *Language and Cognitive Processes, 23,* 1159-1190.

Balota, D. A., Yap, M. J., Cortese, M. J., Hutchison, K. A., Kessler, B., Loftis, B., . . . Treiman, R. (2007). *The English Lexicon Project. Behavior Research Methods, 39,* 445-459.

Boukadi, M., Zouaidi, C., & Wilson, M. A. (2016). Norms for name agreement, familiarity, subjective frequency, and imageability for 348 object names in Tunisian Arabic. *Behavior Research Methods, 48,* 585-599.

Burani, C., Salmaso, D., & Caramazza, A. (1984). Morphological structure and lexical access. *Visible Language, 18,* 348-358.

Burani, C., & Thornton, A. M. (2003). The interplay of root, suffix and whole-word frequency in processing derived words. In R. H. Baayen & R. Schreuder (Eds.), *Morphological Structure in Language Processing* (pp. 157-208). Berlin-New York: Mouton de Gruyter.

Colé, P., Beauvillain, C., & Segui, J. (1989). On the representation and processing of prefixed and suffixed derived words: A differential frequency effect. *Journal of Memory and Language, 28,* 1-13.

Cortese, M. J., & Schock, J. (2013). Imageability and age of acquisition effects in disyllabic word recognition. *The Quarterly Journal of Experimental Psychology, 66,* 946-972.

Durda, K., & Buchanan, L. (2006). *WordMine2.* Retrieved from http://web2.uwindsor.ca/wordmine

Fellbaum, C. (1998). *WordNet: an electronic database.* Cambridge, MA: The MIT Press.

Ford, M. A., Davis, M. H., & Marslen-Wilson, W. D. (2010). Derivational morphology and base morpheme frequency. *Journal of Memory and Language, 63,* 117-130.

Gravetter, F. J., & Wallnau, L. B. (2014). *Essentials of statistics for the behavioral sciences* (8th ed.). Belmont, CA: Wadsworth.

Lehtonen, M., Niska, H., Wande, E., Niemi, J., & Laine, M. (2006). Recognition of inflected words in a morphologically limited language: Frequency effects in monolinguals and bilinguals. *Journal of Psycholinguistic Research, 35,* 121-146.

Luke, S. G., & Christianson, K. (2011). Stem and whole-word frequency effects in the processing of inflected verbs in and out of a sentence context. *Language and Cognitive Processes, 26,* 1173-1192.

Moscoso del Prado Martín, F., Bertram, R., Häikiö, T., Schreuder, R., & Baayen, R. H. (2004). Morphological family size in a morphologically rich language: the case of Finnish compared with Dutch and Hebrew. *Journal of Experimental Psychology: Learning, Memory and Cognition, 30,* 1271-1278.

Sánchez-Gutiérrez, C. H., Mailhot, H., Deacon, S. E., & Wilson, M. A. (2018). MorphoLex: A derivational morphological database for 70,000 English words. *Behavior Research Methods, 50,* 1568-1580.

Taft, M. (1979). Recognition of affixed words and the word frequency effect. *Memory and Cognition, 7,* 263-272.

Yap, M. J., & Balota, D. A. (2009). Visual word recognition of multisyllabic words. *Journal of Memory and Language, 60,* 502-529.

Beyond grammatical category: The role of distributional properties of Italian language in processing nouns and verbs

Daniela Traficante[1]*, Claudio Luzzatti[2], Maria Caterina Silveri[1]

[1] Department of Psychology, Catholic University of Milan, Milan, Italy

[2] Department of Psychology, University of Milan-Bicocca, Milan, Italy
*daniela.traficante@unicatt.it

Abstract

This contribution offers an overview on the main studies carried out on the difference between verb and noun processing in adults and children. This topic will be discussed, in particular, according to the perspective we proposed in collaboration with Cristina Burani, focused on the role of language distributional properties in word processing. This issue has been developed in the last 20 years by assessing differences between words with small inflectional family size, i.e. nouns and adjectives (which in Italian can be inflected in 2-4 word forms), and words with large inflectional family size, i.e. verbs (which can be inflected in up to 50 different word forms). This comparison was made through several tasks (lexical decision, progressive de-masking, reading aloud, a grammatical-class switching task, picture naming) and in different populations, i.e. skilled adult readers, adults with aphasia, adults with Parkinson's Disease, typically developing children and children with dyslexia. In all these studies we found that verb bases are more likely to be processed as base + suffix combination, and are usually associated with slower RTs and lower accuracy than noun bases. Neuroimaging techniques, used in some of these studies, showed that different brain areas are activated in noun and verb processing, as verb recognition seems to involve selection and inhibition mechanisms that are not detected in the case of noun processing. As for nouns, both whole-word and base-word representations seem to be available and their activation is mainly affected by word- and base-frequency, respectively. Overall, data from these studies suggests that the difference between nouns and verbs can depend not only on grammatical, syntactic, and semantic features, but also on the different morphological family size, which might determine difficulty in selection and inhibition processes.

Morphological processing in word recognition

Morphological effects on the word recognition process have been shown in a large amount of experimental data since Taft and Forster's seminal work (1975), but the role of whole word and morpheme representations in the time-course of this process is still under debate (Beyersmann, Coltheart & Castles, 2012). Some models distinguish a very early phase, in which the recognition of a derived or pseudo-derived word begins with a semantically "blind" morpho-orthographic segmentation (Taft, 1994, 2003) and a later phase, in which a morpho-semantic analysis of the morphemic constituents is carried out to assess the likelihood of the specific combination of the identified morphemes. Other authors suggest that all morphologically complex words are first accessed in the lexicon as whole words and morphological decomposition is a post-access process (Giraudo & Grainger, 2001). Finally, the so called 'dual-route models' claim a parallel activation of whole word and morphemic representations in the lexicon (Baayen & Schreuder, 1999; Burani & Caramazza, 1987).

In particular, according to the *Augmented Addressed Morphology* (Burani & Caramazza, 1987; Caramazza, Laudanna & Romani, 1988), morphemic parsing is an auxiliary procedure, that can be applied in front of low-frequency derived words and of pseudo-derived non-words. In this case, the recognition of units larger than single graphemes can offer an advantage in time and accuracy on a letter-by-letter decoding, that, according to Coltheart et al. (2001) *Dual-Route Cascaded Model*, is the procedure needed to decode strings of letters without a corresponding representation in the orthographic input lexicon.

On the other hand, the *Race model* by Schreuder and Baayen (1995), suggests that both morphemic constituents and whole word representations are always activated in front of a morphologically complex word. Two processes would be implemented in parallel: the fast *direct-route*, which leads to the activation of the whole-word representation in the lexicon, and the *parsing-route*, a slower procedure which starts from the decomposition of the string of letters in morphemic units and ends with the assessment of the plausibility of the base-plus-affixes combination. *Race model* is a probabilistic model, in which the balance between costs and benefits determines which route 'wins the race', i.e. which procedure leads to the word recognition faster than the other. Frequency of use is the most important parameter involved in the race: in general, the time-costly parsing-route is slower than the direct-route for high-frequency words, but this might not be true in the case of low-frequency words. In the latter case, when base-word frequency is high, the recognition of morphemic constituents might lead to a faster recognition of the base- plus-affix combination than the access to the low-frequency whole-word representation, should any be available.

Overall, in dual-route models, the kind of representations and the actual process through which a word is accessed are related to specific properties of the word itself: the word, the base, and the affix frequency (token-based measures), the word morphological family size (i.e. the number of words sharing the same base), the affix productivity, and the semantic and phonological transparency of morphemic structure (Laudanna & Burani, 1995).

Evidence from morphologically rich languages, like Finnish (Laine, Vainio & Hyönä, 1999), Hebrew (Deutsch, Frost & Foster, 1998), and Serbo-Croatian (Kostić & Katz, 1987) suggests that also the number of affixes that can be combined with the same base plays a role in the word recognition process. This data supports the model of processing proposed by Niemi, Laine and Tuominen (1994), the *Stem Allomorph/Inflectional Decomposition* (*SAID*), which claims that all morphologically complex words are automatically decomposed. The cost of this parsing procedure make the recognition of polymorphemic words slower than the recognition of simple words.

In Italian, almost all words are morphologically complex: even a base word (e.g. 'libro', book) is the combination of a root ('libr-', book) and a suffix ('-o': singular, masculine). However, there is a huge difference in the number of affixes within words from different grammatical categories. In noun and adjective categories, a root can be combined with 2 through 4 different inflectional affixes, corresponding to gender (masculine, feminine) and number (singular, plural), and -usually- the root frequency is quite balanced within different forms of the same lemma. As for the verb category, more than 50 different inflected forms can be obtained from the combination of the root (e.g., 'pens-', think) with a large number of suffixes corresponding to mode, tense, aspect, person, number (e.g., 'pens-a', 'pens-ava', 'pens-erò', 'pens-erebbero', he/she thinks, thought, I will think, they would think, etc.). In this word category, the root frequency is much higher than the frequency of each inflected form, and it is distributed over a high number of words.

Due to these differences in the morphological family size and in frequency distribution within different root-plus-suffix combinations, it is likely that nouns and verbs are represented differently in the mental lexicon. Italian verbs might be the best candidate to be represented according to the full-parsing view (Pinker & Prince, 1994), which claims that all fully regular inflected words are always accessed by rules. On the other hand, Booij (1993) suggests that plural inflection of nouns is more similar to derivation than to inflection process, as the pluralization of a noun changes the meaning in a way in which the verb inflection does not, leading to a new concept. Evidence from Dutch (Baayen, Dijkstra, & Schreuder, 1997) and from Italian (Baayen, Burani, & Schreuder, 1997) singular- and plural-dominant

146

nouns (e.g. nose – singular dominant; teeth – plural dominant) is consistent with this view, because it showed that also plural nouns may be accessed as full-forms.

Colombo and Burani (2002), contrasting verbs and nouns in a lexical decision task (LDT), found that root frequency affects RTs for verbs more than for nouns, and that latencies for verbs are longer than for nouns. Similarly, Traficante and Burani (2003) and Traficante, Barca and Burani (2004) found that verbs were associated with slower RTs both in LDT and in a Progressive De-masking Task (PDT) in comparison to adjectives, even though the two lists of words were matched by length, frequency and imageability. Post-hoc correlations showed that root frequency affects only latency to verbs, whereas whole-word frequency is the only variable affecting responses to adjectives.

In this vein, Laudanna, Voghera and Gazzellini (2002), using a priming paradigm, found evidence that noun- and verb-bases might be stored and accessed in different ways. The authors suggested that noun bases are likely to be represented, in the orthographic input lexicon, both as a whole word and in decomposed form, whereas verb bases are likely to be represented only in decomposed form.

More recently, Traficante, Marelli, Luzzatti and Burani (2014) presented primary-school children with derived nouns in a reading aloud task. Their results confirm that nouns derived from verb bases (e.g., 'departure') are processed more slowly than nouns derived from noun bases (e.g., 'humorist'). Moreover, the base frequency and the whole-word frequency (that can be interpreted as the probability of base-plus-suffix combination) affect nouns derived from a verb, but does not influence nouns derived from a noun.

To understand the interplay between base and whole-word representations better, Traficante, Marelli and Luzzatti (2018) presented primary-school children with the same stimuli used in the previous work, embedded in sentences, and recorded children's eye movements while reading the stimuli aloud. Results indicate that, for nouns derived from noun base, the higher the frequency of the base is, the faster the first fixation duration, but this advantage can be contrasted by the competition with whole-word representation. This competition does not occur in gaze duration. These results are consistent with data proving that base recognition is a head-start only when word frequency is low (Marcolini, Traficante, Zoccolotti, & Burani, 2011).

On the contrary, for nouns derived from a verb, the base exerts an inhibitory effect on first fixation, whereas whole-word frequency influences gaze duration in the expected direction: the higher the frequency is, the shorter the gaze duration. The inhibitory effect of base frequency on first fixation is not consistent with data from reading aloud single words, and it has been considered as a consequence of the syntactic structure of the sentence within which the target word was embedded. From a syntactic point of view, a noun is expected in the target place, whereas a verb base is activated at first. In this condition, the reader realizes that the actual target must be the noun derived from that verb base, and a derivation process is triggered, in order to select the target noun within the large morphological family of that verb base. As Marangolo, Piras, Galati, and Burani (2006) study showed (see below), this is a very difficult task, and long first fixation duration observed in reading noun derived from a verb base can be indicative of the effort required in the derivation process.

Irrespective of the (inhibitory or facilitative) direction of base frequency effect on first fixation duration, data from eye-movement recording clearly shows that the first morphemic constituent plays a relevant role in early processing of written words. Burani, Marcolini, Traficante and Zoccolotti (2018) found that the longer this constituent is, the higher its effect in reading aloud single words. However, this result is proved for young skilled readers, but not for poor readers, who always gain advantage from the opportunity of detecting a base word at the beginning of a string of letters, irrespective of its length.

Morphological processing in word retrieval

Differences in processing verbs and nouns have been observed not only in word recognition task (word naming, lexical decision task), but also in word retrieval after left-hemisphere brain damage, and this evidence supports the hypothesis that verbs and nouns must be represented in a different way at a level of processing shared by both input and output procedures (Crepaldi et al., 2006). Levelt, Roelofs and Meyer (1999) proposed a model of lexical access where representations are organized in a two-layers system: the *lemma level*, a central lexical-syntactic store where grammatical (gender for nouns, thematic structure for verbs) and conceptual information is represented in a modality-independent way; the peripheral *lexeme level*, where orthographical and phonological word form is encoded.

It is under debate at which level of processing noun-verb dissociation in aphasia can be referred to, but some evidence on this issue can be drawn from studies of brain-damaged patients. In neuropsychological literature, there are several reports on people with neurological damage, who present dissociable deficit for nouns and verbs. Usually this evidence comes from word retrieval tasks (e.g., picture naming) and there is a wide convergence on the conclusion that different classes of words can be supported by segregated neural substrates (Basso et al., 1990; Caramazza & Hillis, 1991; Chen & Bates, 1998; Daniele, Giustolisi, Silveri, Colosimo, & Gainotti, 1994; Daniele, Silveri, Giustolisi, & Gainotti, 1993; Luzzatti et al., 2002).

A complex pattern of results were found by Caramazza and Hillis, who observed double dissociations between nouns and verbs in different tasks (oral vs. written naming): some patients managed to produce verbs in picture naming, but not to write them down (Caramazza & Hillis, 1991); other patients were impaired in producing nouns orally, but had greater impairment either in understanding written verbs or in writing verb forms (Hillis & Caramazza, 1995). These results led the authors to suggest damage to the phonological and orthographical representations of words, at the *lexeme level* (Levelt et al., 1999).

In contrast, Berndt, Mitchum, Haendiges, and Sandson (1997), after considering data from 10 aphasic patients, who showed a variety of different impairments, suggested that noun-verb dissociation must involve a lexical device, either at the level of orthographic-phonological representations (lexeme level) or at the level of a lexical-syntactic amodal device (lemma level).

Bird and coll. (Bird, Howard, & Franklin, 2000; Bird, Ralph, Patterson, & Hodges, 2000), who found predominant V impairment in aphasic patients, referred N-V dissociation to semantic differences. Nouns are more concrete and imageable than verbs, that are more abstract, and characterized by movement/motor attributes. In particular, imageability is assumed to be the most important predictor of a patient's ability to retrieve words (Bates, Burani, D'Amico & Barca, 2001; Luzzatti et al., 2002) and this effect is likely to be increased by picture naming, the task most used to assess patients' lexical production. The 'Right Hemisphere Hypothesis' (RHH) suggested by Coltheart (1987, 2000) claims that, after full left-hemisphere lexical damage, a naming system located in the right hemisphere can be involved in word production. The imageability effect would come from the features of the lexical knowledge of the right hemisphere, which would be limited to high-frequency, concrete words. According to this hypothesis, Crepaldi et al. (2006) suggested that verb impairment might be explained as a consequence of the emergence of right-hemisphere lexical abilities after large left-hemisphere perisylvian damage.

To test the role of imageability in determining the N-V dissociation, Crepaldi et al. (2006) developed a specific task, the Noun and Verb Retrieval in a Sentence Context (NVR-SC), in which the same image was associated with two sentences that share the same meaning, but differed in syntactic structure: in the N-to-V condition. The first sentence contained a noun (e.g. Mary likes *conversation*) and the second sentence presented a gap, that the respondent had to fill in by producing the corresponding verb (e.g., Mary likes _____ [to converse]); in the V-to-N condition the sentences were in the inverted sequence. In this task, target nouns and verbs were perfectly matched for imageability, as all nouns were derived from the corresponding verbs, or vice versa. Data from NVR-SC task were compared to the patients' performance in a classical picture naming task. Grammatical class effect emerged only in the latter task (nouns: 73% of correct answers vs. verbs: 35%), whereas in NVR-SC no difference emerged between nouns and verbs (nouns: 58% vs. verbs: 56%).

These results prove a strong association between verb impairment and imageability effect. Such effect, according to Luzzatti and Chierchia (2002), might be the consequence of a compensatory strategy, that patients are prone to apply in front to the difficulty in accessing the argument structure of verbs at the lemma level (Levelt et al., 1999). When the access to this structure is impaired, due to an extended lesion in the left hemisphere, a reconstruction of the argument structure through a mental image of the action (arguably through the right hemisphere) can be helpful in verb retrieval. According to this hypothesis, the NVR-SC task can increase the percentage of correct answers for verbs (56% vs. 35% in picture naming) as it offers a sentence frame in which argument structure is already provided. This explanation referred to verb argument structure can also account for the lack of similar advantage for nouns in NVR-SC, in comparison to a picture naming task. In fact, for nouns, argument structure is less essential than for verbs, and imageability of nouns presented in picture naming is higher than imageability of nouns in NVR-SC. This unbalance between the items of the two tasks can account for the significant difference in correct answers for nouns (73% in the picture naming vs. 58% in the NVR-SC task).

However, further results from multivariate logistic regression analysis showed that in 3 out of 16 patients grammatical category remained significant, even though imageability and word frequency effects were partialled out. These data suggest an additional verb-specific damage, which has been referred by the authors to the lemma level, according to Levelt et al. (1999) model.

Some clues of peculiarity of verb processing come also from the study of morphological task in Parkinson's Disease (PD) patients. These patients are usually impaired in verb production and such deficit has been interpreted as a disexecutive disorder (Dubois e Pillon, 1996; Green et al., 2002), i.e. a deficit of selection processing in word production, due to corticostriatal damage. Silveri et al. (2018) presented 14 PD patients and 14 healthy controls with the stimuli and the morphological tasks used by Marangolo et al. (2003). The participants read the input stimuli on a computer screen and produced their responses orally. Four conditions were administered: 1) derivation of a noun from a verb (VN: 'observation' from 'to observe'); 2) derivation of a noun from an adjective (AN: 'kindness' from 'kind'); 3) generation of a verb from a noun (NV: 'to fail' from 'failure'); 4) generation of an adjective from noun (NA: 'fresh' from 'freshness'). Derivation tasks were more difficult than generation tasks for all participants, but PD patients were significantly more impaired than controls in deriving nouns from verbs. This data was interpreted as the consequence of impaired selection processes among competitors, which can be considered typical of PD patients, as an expression of the dysexecutive disorder.

Neuro-anatomical correlates of verb and noun processing

Neuroimaging studies showed a quite stable association between lesions of the left temporal lobe and impaired production of nouns (Damasio, Tranel, Grabowskia, Adolphsa, & Damasio, 2004; Daniele et al., 1994; Silveri & Di Betta, 1997). On the other hand, deficit in the production of verb forms turned out to be mostly associated to a wide range of lesions in the fronto-parietal regions (Silveri, Perri, & Cappa, 2003; Silveri, Salvigni, Cappa, Della Vedova, & Puopolo, 2003) and connected subcortical structures, in particular the basal ganglia (Bertella, Albani, Greco, Priani et al., 2002; Bocanegra et al., 2015; Cotelli, Borroni, Manenti, Ginex et al., 2007; Fernandino et al., 2013).

Data provided by Aggujaro et al. (2006) showed a more complex pattern of results. It is, to the best of our knowledge, the first group research on the neuro-anatomical correlates of N-V dissociation. Participants were 20 patients with aphasia. Three out of 5 patients with predominant noun impairment had lesions in the medial part of the middle and inferior left temporal gyri, as expected from the previous studies, but 2 patients had left inferior and mesial occipito-temporal damage. Twelve out of 15 patients with prevalent impairment of verbs showed damage in different areas, which have been clustered in three major subsets: 1) Four non-fluent aphasic patients showed large left hemisphere lesions extending to the frontal and temporal territory, damaging all the left hemisphere language areas; 2) Four fluent aphasic patients had lesions in a left posterior temporal and inferior parietal area; 3) Four patients had damage only in the insular cortex and/or in the underlying subcortical structures. The remaining patients had a large deep-seated lesion associated to a small cortical parietal lesion or a temporal mesial left-hemisphere lesion. None of the patients with verb impairment had an isolated pre-frontal lesion.

In order to get clues on the neural correlates of morphological processing on different grammatical classes, Marangolo et al. (2006) presented 10 healthy adults with three different tasks: 1) derivation from verb to noun (e.g., 'observation' from 'to observe'), from adjective to noun (e.g., 'kindness' from 'kind'), and generation from noun to verb (e.g., 'to fail' from 'failure'); 2) inflection (past participle for verbs, plural forms for nouns and adjectives; 3) word repetition (used as a baseline for fMRI).

Evidence from fMRI showed that derivational processing (noun-from-verb; noun-from-adjective) activates a network including the ventrolateral frontal cortex (bilaterally when the noun was produced from a verb and limited to the left hemisphere when the noun was produced from an adjective) and the inferior parietal lobule (IPL) of both the hemisphere. The frontal activation was bilateral for noun derivation from verb, but was limited to the left hemisphere for noun derivation from adjective. Most of the left frontal activation was located in the pars triangularis and pars opercularis of the inferior frontal gyrus (IFG). Verb-from-noun generation task only activated the left IFG.

151

As for verb inflection, activation of the IFG and IPL in the left hemisphere were observed, but at a much smaller extent than for the derivation process. This activation has been interpreted as a correlate of processes of decomposition of a verb into its constituent morphemes. Noun inflection activated only a small region in the left insula and this result was interpreted as evidence that noun plurals are subject to lesser segmentation than verb inflected forms. Finally, adjective inflection did not activate either the IFG or the IPL. This result suggests that adjectives are less subject to morphological decomposition than verbs and their inflection has a primarily semantic role in sentence interpretation. This fMRI data is fully consistent with behavioral results from word recognition tasks (Colombo & Burani, 2002; Traficante, Barca, & Burani, 2004; Traficante & Burani, 2003) and prove that, in Italian, verb processing involves mechanisms and representations that differ from those involved in noun and adjective processing.

Berlingeri et al. (2008) presented 12 healthy Italian undergraduate students with a task that is similar to the task employed by Marangolo et al. (2006) (Grammatical Class Switching Task), in which participants had either to read a noun and retrieve the corresponding verb (generation task), or to read a verb and retrieve the corresponding noun (derivation task). Analyses from fMRI showed verb-related activation of a dorsal premotor and posterior parietal network, whereas no brain area was consistently associated with nouns.

Recently, the same task was used by Di Tella et al. (2018), to study the association between performance on morphological tasks, and cortical and subcortical measures of thickness, obtained through structural MRI examination. The authors compared patients with Parkinson's Disease with nigrostriatal hypofunctionality in the left hemisphere (PD-LH), patients with Parkinson's Disease with nigrostriatal hypofunctionality in the right hemisphere (PD-RH) and a control group. Results of the previous study (Silveri et al., 2018) were confirmed: the lowest percentage of correct responses was observed in the noun-from-verb condition (that is, the most difficult one), but only in the PD group with major left-hemisphere damage (PD-LH). Moreover, only in this population a significant negative correlation emerged between cortical thickness of the left IFG and accuracy, not only for noun but also for verb production, that is, whenever the production of a word is required independently of its grammatical class. This result has been taken as evidence of the interaction between executive resources and language in the left IFG.

Conclusions

Both neuroimaging studies and behavioral research suggest that the large morphological family size of Italian verbs leads to word parsing into morphemic constituents. This process can gain advantage from high frequency of the root, but it is time-costly (see *Race model* by Baayen & Schreuder, 1999), since both verb recognition and verb production require the selection of the right root-plus-suffix combination among several (about 50) alternatives, with inhibition of competitors. On the contrary, nouns and adjectives are likely to be represented and processed as whole words, without the involvement of parsing and combinatorial processes.

Due to its combinatorial and distributional features, the Italian language offers a good opportunity to study the role of morphological family size, i.e. the number of inflected and derived forms sharing the same word base, on word recognition and word production. In the present work a brief review of research contrasting verbs (large morphological family size) and nouns (small morphological family size) has been presented, both on adults and on children in Italian.

Studies carried out with neuroimaging techniques in different populations (healthy adults, patients with Parkinson's Disease, Aphasic patients) offered useful clues to understand the neural networks involved in verb and noun processing better, and contributed in shedding light on the selection and inhibition processes involved in the access to words derived from verb base.

References

Aggujaro, S., Crepaldi, D., Pistarini, C., Taricco, M., & Luzzatti, C. (2006). Neuroanatomical correlates of the impaired retrieval of verbs and nouns: Interaction of grammatical class, imageability and actionality. *Journal of Neurolinguistics, 19*, 175-194.

Baayen, R.H., Burani, C., & Schreuder, R. (1997). Effects of semantic markedness in the processing of regular nominal singulars and plurals in Italian. In Booij G. and van Marle J. (Eds.), *Yearbook of Morphology* 1996 (pp. 13-33). Dordrecht: Kluver Academic Publisher.

Baayen, R. H., Dijkstra, T., & Schreuder, R. (1997). Singulars and plurals in Dutch: Evidence for a parallel dual-route model. *Journal of Memory and Language, 37*, 94-117.

Baayen, R.H., & Schreuder, R. (1999). War and peace: Morphemes and full forms in a non-interactive activation parallel dual route model. *Brain and Language, 68*, 27-32.

Basso A., Razzano C., Faglioni P. & Zanobio E. (1990). Confrontation naming, picture description and action naming in aphasic patients. *Aphasiology, 4*, 185-95.

Bates, E., Burani, C., D'Amico, S., & Barca, L. (2001). Word reading and picture naming in Italian. *Memory & Cognition, 29*, 986-999.

Berlingeri, M., Crepaldi, D., Roberti, R. Scialfa, G., Luzzatti, C., & Paulesu, E. (2008). Nouns and verbs in the brain: Grammatical class and task specific effects as revealed by fMRI. *Cognitive Neuropsychology, 25*, 528-558.

Berndt, R. S., Mitchum, C. C., Haendiges, A. N., & Sandson, J. (1997). Verb retrieval in aphasia. 1. Characterizing single word impairments. *Brain and Language, 56,* 68-106.

Bertella, L., Albani, G., Greco, E., Priano, L., Mauro, A., Marchi, S.... & Semenza, C. (2002). Noun-verb dissociation in Parkinson's disease. *Brain and Cognition, 48,* 277-280.

Beyersmann, E., Coltheart, M., & Castles, A. (2012). Parallel processing of whole words and morphemes in visual word recognition. *The Quarterly Journal of Experimental Psychology, 65,* 1798-1819.

Bird, H., Howard, D., & Franklin, S. (2000). Why is a verb like an inanimate object? Grammatical category and semantic category deficits. *Brain and Language, 72,* 246-309.

Bird, H., Ralph, M. A. L., Patterson, K., & Hodges, J. R. (2000). The rise and fall of frequency and imageability: Noun and verb production in semantic dementia. *Brain and Language, 73,* 17-49.

Bocanegra, Y., García, A. M., Pineda, D., Buriticá, O., Villegas, A., Lopera, F. ... & Trujillo, N. (2015). Syntax, action verbs, action semantics, and object semantics in Parkinson's disease: Dissociability, progression, and executive influences. *Cortex, 69,* 237-254.

Booij, G.E. (1993). Against Split Morphology. In G.E. Booij and J. van Marle (Eds.), *Yearbook of Morphology 1993* (pp. 27-49). Dordrecht: Kluwer Academic Publishers.

Burani, C., & Caramazza, A. (1987). Representation and processing of derived words. *Language and Cognitive Processes, 2,* 217-227.

Burani, C., Marcolini, S., Traficante, D., & Zoccolotti, P. (2018). Reading derived words by Italian children with and without dyslexia: The effect of root length. *Frontiers in Psychology, 9:* 647.

Caramazza, A., & Hillis, A. (1991). Lexical organization of nouns and verbs in the brain. *Nature, 349,* 788-790.

Caramazza, A., Laudanna, A., & Romani, C. (1988). Lexical access and inflectional morphology. *Cognition, 28,* 297-332.

Chen S., & Bates E. (1998). The dissociation between nouns and verbs in Broca's and Wernicke's aphasia: findings from Chinese. *Aphasiology, 12,* 5-36.

Colombo, L., & Burani, C. (2002). The influence of age of acquisition, root frequency, and context availability in processing nouns and verbs. *Brain and Language, 81,* 398-411.

Coltheart, M. (1987). Deep dyslexia: A right hemisphere hypothesis. In M. Coltheart and K. Patterson (Eds.), *Deep dyslexia* (2nd ed., pp. 326–380). London: Routledge & Kegan Paul.

Coltheart, M. (2000). Deep dyslexia is right-hemisphere reading. *Brain and Language, 71,* 299-309.

Coltheart, M., Rastle, K., Perry, C., Langdon, R., & Ziegler, J. (2001). DRC: A dual route cascaded model of visual word recognition and reading aloud. *Psychological Review, 108*, 204-256.

Cotelli, M., Borroni, B., Manenti, R., Ginex, V., Calabria, M., Moro, A. ... & Padovani, A. (2007). Universal grammar in the frontotemporal dementia spectrum: Evidence of a selective disorder in the corticobasal degeneration syndrome. *Neuropsychologia, 45*, 3015-3023.

Crepaldi, D., Aggujaro, S., Arduino, L. S., Zonca, G., Ghirardi, G., Inzaghi, M. G., ... & Luzzatti, C. (2006). Noun–verb dissociation in aphasia: The role of imageability and functional locus of the lesion. *Neuropsychologia, 44*, 73-89.

Damasio, H., Tranel, D., Grabowski, T., Adolphs, R., & Damasio, A. (2004). Neural systems behind word and concept retrieval. *Cognition, 92*, 179-229.

Daniele, A., Silveri, M. C., Giustolisi, L., & Gainotti, C. (1993). Category-specific deficits for grammatical classes of words: Evidence for possible anatomical correlates. *The Italian Journal of Neurological Sciences, 14*, 87-94.

Daniele, A., Giustolisi, L., Silveri, M.C., Colosimo, C., & Gainotti, G. (1994). Evidence for a possible neuroanatomical basis for lexical processing of nouns and verbs. *Neuropsychologia, 32*, 1325-1341.

Deutsch, A., Frost, R., & Forster, K.I. (1998). Verbs and nouns are organized and accessed differently in the mental lexicon: Evidence from Hebrew. *Journal of Experimental Psychology: Learning, Memory and Cognition, 24*, 1238-1255.

Di Tella, S., Baglio, F., Cabinio, M., Nemni, R., Traficante, D., & Silveri, M. C. (2018). Selection processing in noun and verb production in left- and right-sided Parkinson's Disease patients. *Frontiers in Psychology, 9*, n. 1241.

Dubois, B., & Pillon, B. (1996). Cognitive deficits in Parkinson's disease. *Journal of Neurology, 244*, 2-8.

Fernandino, L., Conant, L. L., Binder, J.R., Blindauer, K., Hiner, B., Spangler, K., & Desai, R. H. (2013). Where is the action? Action sentence processing in Parkinson's disease. *Neuropsychologia, 51*, 1510-1517.

Giraudo, H., & Grainger, J. (2001). Priming complex words: Evidence for supralexical representation of morphology. *Psychonomic Bulletin & Review, 8*, 127-131.

Green, J., McDonald, W. M., Vitek, J. L., Evatt, M., Freeman, A., Haber, M., ... & DeLong, M. R. (2002). Cognitive impairments in advanced PD without dementia. *Neurology, 59*, 1320-1324.

Hillis, A., & Carammazza, A. (1995). Representation of grammatical knowledge in the brain. *Journal of Cognitive Neuroscience, 7*, 369-407.

Kostić, A., & Katz, L. (1987). Processing differences between nouns, adjectives, and verbs. *Psychological Research, 49*, 229-236.

Laine, M., Vainio, S., & Hyönä, J. (1999). Lexical access routes to nouns in a morphologically rich language. *Journal of Memory and Language, 40*, 109-135.

Laudanna, A., & Burani, C. (1995). Distributional properties of derivational affixes: Implications for processing. In L.B. Feldman (Ed.), *Morphological Aspects of*

Language Processing: Cross-Linguistic Perspectives (pp. 345-364). Hillsdale (NJ): Lawrence Erlbaum Associates.

Laudanna, A., Voghera, M., & Gazzellini, S. (2002). Lexical representations of written nouns and verbs in Italian. *Brain and Language, 81*, 25-263.

Levelt, W. J., Roelofs, A., & Meyer, A. S. (1999). A theory of lexical access in speech production. *Behavioral and Brain Sciences, 22*, 1-38.

Luzzatti, C., & Chierchia, G. (2002). On the nature of selective deficit involving nouns and verbs. *Rivista di Linguistica, 14*, 43-71.

Luzzatti, C., Raggi, R., Zonca, G., Pistarini, C., Contardi, A., & Pinna, G. D. (2002). Verb–noun double dissociation in aphasic lexical impairments: The role of word frequency and imageability. *Brain and Language, 81*, 432-444.

Marangolo, P., Incoccia, C., Pizzamiglio, L., Sabatini, U., Castriota-Scanderbeg, A., & Burani, C. (2003). The right hemisphere involvement in the processing of morphologically derived words. *Journal of Cognitive Neuroscience, 15*, 364-371.

Marangolo, P., Piras, F., Galati, G., & Burani, C. (2006). Functional anatomy of derivational morphology. *Cortex, 42*, 1093-1106.

Marcolini, S., Traficante, D., Zoccolotti, P., & Burani, C. (2011). Word frequency modulates morpheme-based reading in poor and skilled Italian readers. *Applied Psycholinguistics, 32*, 513-532.

Niemi, J., Laine, M., & Tuominen, J. (1994). Cognitive morphology in Finnish: Foundations of a new model. *Language and Cognitive Processes, 9*, 423-446.

Pinker, S., & Prince, A. (1994). Regular and irregular morphology and the psychological status of rules of grammar. In S.D. Lima, R.L. Corrigan, and J.K. Iverson (Eds.), *The reality of linguistic rules* (pp. 321-351). Philadelphia (PA): John Benjamins.

Schreuder, R., & Baayen, R.H. (1995). Modeling morphological processing. In L.B. Feldman (Ed.), *Morphological Aspects of Language Processing* (pp. 131-154). Hillsdale (NJ): Lawrence Erlbaum.

Silveri, M. C., & Di Betta, A. M. (1997). Noun–verb dissociations in brain damaged patients: further evidence. *Neurocase, 3*, 477-488.

Silveri, M. C., Perri, R., & Cappa, A. (2003). Grammatical class effects in brain-damaged patients: functional locus of nouns and verb deficits. *Brain and Language, 85*, 49-66.

Silveri, M. C., Salvigni, B. L., Cappa, A., Della Vedova, C., & Puopolo, M. (2003). Impairment of verb processing in frontal variant-frontotemporal dementia: a dysexecutive symptom. *Dementia and Geriatric Cognitive Disorders, 16*, 296-300.

Silveri, M. C., Traficante, D., Lo Monaco, M. R., Iori, L., Sarchioni, F., & Burani, C. (2018). Word selection processing in Parkinson's disease: When nouns are more difficult than verbs. *Cortex, 100*, 8-20.

Taft, M. (1994). Interactive-activation as a framework for understanding

156

morphological processing. *Language and Cognitive Processes, 9,* 271-294.

Taft, M. (2003). Morphological representation as a correlation between form and meaning. In E. Assink and D. Sandra (Eds.), *Reading complex words* (pp. 113-137). Amsterdam (NL): Kluwer Academic Publishers.

Taft, M., & Forster, K. I. (1975). Lexical storage and retrieval of prefixed words. *Journal of Verbal Learning and Verbal Behavior, 14,* 638-647.

Traficante, D., Barca, L., & Burani, C. (2004). Accesso lessicale e lettura ad alta voce: il ruolo delle componenti morfologiche delle parole. *Giornale Italiano di Psicologia, 31,* 821-838.

Traficante, D., & Burani, C. (2003). Visual processing of Italian verbs and adjectives: The role of inflectional family size. In R. H. Baayen and R. Schreuder (Eds.), *Morphological Structure in Language Processing* (pp. 45-64). Berlin: Mouton de Gruyter.

Traficante, D., Marelli, M., & Luzzatti, C. (2018). Effects of reading proficiency and of base and whole-word frequency on reading noun- and verb-derived words: an eye-tracking study in Italian primary school children. *Frontiers in Psychology, 9:* 2335.

Traficante, D., Marelli, M., Luzzatti, C., & Burani, C. (2014). Influence of verb and noun bases on reading aloud derived nouns: Evidence from children with good and poor reading skills. *Reading and Writing, 27,* 1303-1326.

Theories of morphological analysis: a mini-review

Anna M. Thornton[1]*

[1]Università dell'Aquila

*annamaria.thornton@univaq.it

Abstract.
This paper presents a short review of arguments in favor of inferential-realizational, rule-based, Word and Paradigm theories of morphology, and against lexical-incremental, morpheme-based, Item and Arrangement theories. Morpheme-based theories are assumed by most linguists who do not specialize in morphology, while most contemporary morphologists favor Word and Paradigm theories, developed particularly in work by Peter H. Matthews, Stephen R. Anderson, Mark Aronoff and Gregory T. Stump.

I met Cristina Burani when I had just completed my PhD in Linguistics (University of Pisa, 1989), working on derivational morphology. Cristina practically adopted me, and throughout the Nineties I cooperated with her and other colleagues at the Istituto di Psicologia del CNR in Rome in a number of experiments devoted to assessing the role of derivational affixes in processing derived words. The assumptions under which I operated at the time were those that I have now come to recognize as typical of so-called "morpheme-based" (Anderson, 2017) or "lexical-incremental" (Stump, 2001) or "Item and arrangement" (Hockett, 1954) models of morphology. However, at that time, I had no awareness that such models represented only one of a number of possible options available to understand the way in which morphologically complex words (both inflected and derived) come about. I took for granted that morphologically complex words were composed of two or more (usually bound) entities, called morphemes, and I spent years happily (well, more or less happily...) counting the frequency of specific roots and affixes in the then available corpora of written Italian.[1] At the end of the Nineties, my cooperation with Cristina came to an end, for a number of partially unrelated reasons: increasing teaching and administrative responsibilities at the University of L'Aquila left me little time for research, and in this little time I was becoming more and more interested in inflectional morphology, rather than in derivation. Applying a morpheme-based model to the analysis of Italian inflection proved extremely frustrating (see my desperate attempt in Thornton, 1999), and forced me to look beyond what I had tacitly and unquestioningly assumed so far, to understand how the morphology of a language works.

In this mini-review I will try to share what little wisdom I have acquired in the process of exploring alternative models of morphological analysis, and argue in favor of Word and Paradigm, rule-based, inferential-realizational theories of morphology.

Several characterizations of models of morphological analysis are available. The earliest outline of possible alternatives was offered by Hockett (1954), who recognized three models, called Item and Arrangement (IA), Item and Process (IP), and Word and Paradigm (WP). Anderson (2015) reduces the alternative to two options, which he calls morpheme-based and rule-based. Stump (2001) offers a quadripartite typology, which will be illustrated below.

A very clear statement of how an IA, morpheme-based model operates is given by Bloch (1947, pp.399-400): "Any sentence, phrase or complex word can be

---

[1]

The importance of having reliable frequency data was felt so much that we embarked in creating a new resource, CoLFIS (Bertinetto et al., 2005).

described as consisting of such-and-such morphemes in such-and-such an order. [...] The preterit form *waited* [...] consists of two morphemes, /weyt/ and /ed/ , occurring in that order. The meaning of the first morpheme is a particular action that we need not specifically describe here; that of the second is 'past time' or the like".

Let us note in passing, but not dwell upon, the fact that the IA model is characterized as valid for both morphology and syntax, and let us concentrate on morphology. Examples of complex words that seem to comply with such a model are often cited: Anderson (2015) uses *unavoidable* 'not possible to avoid', where *un-* expresses 'not', *-able* expresses 'possible' and *avoid* expresses 'to avoid', leaving no part of both form and meaning of the complex word unaccounted for; Carstairs-McCarthy (2005) cites *unhelpfulness*; Burani (2006) cites Italian *deindustrializzazione* 'deindustrialization'. Maybe it is not an accident that these examples are all of derived lexemes, not of inflected forms of lexemes. CarstairsMcCarthy (2005, p.20) explicitly observes that in dealing with derivational morphology there is "less temptation to deviate from the rough-and-ready sense of 'morpheme'", i.e., that of a Saussurean sign, which unites a signifier and a *signifié* and is not further analyzable in smaller units that are still signs. In inflectional morphology, instead, although one can find examples of inflectional forms that lend themselves to an IA analysis without causing immediate difficulties (such as Bloch's example *waited*), it is very common to find forms that defy an IA analysis completely, such as Italian *è* 'be.PRS.IPFV.IND.3SG' and *fu* 'be.PST.PFV.IND.3SG', or English *am* 'be.PRS.IND.1SG', to cite just three of the innumerable examples in which it is impossible to segment the signifier of a word form in meaningful units each of which corresponds to a part of the word's meaning.

This problem has led most contemporary morphologists (although not most contemporary linguists whose main field is not morphology, unfortunately) to abandon morpheme-based theories of word structure and embrace theories that fall within the model that Hockett called Word and Paradigm. Stump (2001, pp.1-12; see also Stump, 2016, pp.8-30) gives a very useful presentation of criteria that can be used to classify morphological theories, and distinguishes four families of theories. He first contrasts lexical and inferential theories (and credits Andrew Spencer for suggesting the term "inferential", cf. Stump, 2001, p.277, fn. 2). These two kinds of theories give different answers to the fundamental question of what is the nature of the association between an inflected word's morphosyntactic properties (also called grammatical feature values: that is, things such as **TENSE**: PRESENT, **NUMBER**: SINGULAR, etc.) and the signifier of the word form that has such properties. In lexical theories, the association between a set of properties and a signifier is listed in the lexicon, much in the same way as the association between a "lexical morpheme" and its meaning is: so, e.g., English *-s* would be the signifier of a morpheme whose meaning is '3SG.PRS.IND'; in inferential theories, instead, rules express the association between a lexeme and its inflected forms, with no requirement that specific morphosyntactic properties be linked to specific bits of the signifier of a word form. The distinction between lexical and inferential theories cross-cuts another distinction, between incremental and realizational theories. In incremental theories, there is a requirement that inflected words acquire certain morphosyntactic properties "only as a concomitant of acquiring the inflectional exponents of those properties" (Stump, 2001, p.2), while in realizational theories "a word's association with a particular set of morphosyntactic properties licenses the introduction of those properties' inflectional exponents" (Stump, 2001, p.2). Stump strongly advocates in favor of inferential-realizational theories, and presents a number of arguments that support such theories against alternative ones.

A first argument comes from the existence of extended and overlapping exponence, i.e., the fact that sometimes a given property is not expressed by just a single exponents within a word form, and often an exponent expresses more than one property: a well-known illustration of the intricacies of such multiple exponence relations has been offered by Matthews (1991, pp.168-184), who took as an examples the ancient Greek form *elelýkete* 'you had unfastened'. Matthews (1975) exemplifies with Italian conditional forms: in a form such as *canterebbero* 'they would sing' ('sing.PRS.COND.3PL'), the property '3rd person' is expressed both by the portion *-bbe-* (which appears only in 3rd person forms of the present conditional) and by the portion *-ro*, which appears only in 3rd person plural forms of Italian verbs (cf. *videro* 'they saw'); in their turn, *-bbe-* is also one of the exponents of the property 'conditional', and this is a case of overlapping exponence; *-ro* is also an exponent of the property 'plural', and this is a case of cumulative exponence.[2]

---

[2] Matthews (1991) distinguishes several kinds of exponence relations, i.e. relations between portions of the form of an inflected form and its meaning: **simple** exponence is the one-to-one relation between a portion of form and a property of meaning (e.g. *-ed* and 'past' in *waited*); **cumulative** exponence is the case (very familiar from Indo-European languages) in which a single element of form expresses more than one property and these properties are always expressed together in the language (e.g. *-o* in Italian *buono* 'good.M.SG' expresses cumulatively **GENDER**: MASCULINE and **NUMBER**: SINGULAR); **overlapping** exponence is the situation in which a certain property is expressed by some exponent which also expresses other properties, but cumulation of the expression of the two properties is not general in the language (e.g. in ancient Greek the endings *-te* and *-st^{h}e* cumulatively express **PERSON**: SECOND and **NUMBER**: PLURAL, and in addition *-te* expresses **VOICE**: ACTIVE and *-st^{h}e* **VOICE**: PASSIVE, but voice is expressed also by other means in Greek verb forms, so its expression is not always cumulated with the expression of person and number, but sometimes overlaps with it); **extended** exponence is the case in which more than one element of form expresses a single property (e.g. in English *sold* 'sell.PST' the

Lexical and incremental theories would have problems justifying the appearance of the same meaning twice, carried (individually or in a cumulative or overlapping manner) by two different exponents within a word, while inferential and realizational theories do not imply a ban on extended or overlapping exponence.

A second strong argument against lexical theories is the existence of cases of underdetermination, a situation in which "the word form realizing a particular morphosyntactic property set does not contain exponents of all the properties in that set" (Stump, 2016, p.17). Stump's most readily understandable example is English *cut*, which can correspond to several forms of the paradigm of the verb 'cut' (the infinitive, a non-3rd person present indicative form, a past tense form, a past participle form, the imperative...). Lexical-incremental, morpheme-based, IA theories have faced this problem by recurring to the theoretical construct of the zero-morph: if we need a piece of signifier to express a certain property in a word form, but this piece is not audible, we can posit a morpheme which has nothing ('zero') as its signifier, but carries the relevant property as its meaning. This solution has been widely adopted in the description of inflection by American structuralists in the Forties, and just as widely criticized by modern morphologists who adhere to rule-based, inferential-realizational, WP models of morphological analysis. Let us see how Bloch (1947) introduces a zero morph within his morpheme-based, IA analysis of English verb inflection. After analyzing *waited* 'wait.PST' as the concatenation of the two morphemes *wait-* 'wait' and *-ed* 'past', Bloch addresses the issue of how to analyze other past tense forms, such as *passed*, *lived, put* and *took.* Comparison of *waited, passed* and *lived* reveals "three phonemic shapes of the preterit suffix. In general, the choice among them depends on the last phoneme in the base"; *put* "contains an additional shape of the same suffix, namely zero" (Bloch, 1947, p.402). Bloch's analysis of *put* 'put.PST' is then that this form is made up of two morphemes, *put* 'put' and zero 'past'. It would appear that at the cost of accepting that some morphemes can have zero as their signifier, a morpheme-based model of morphological analysis can be saved. But more problems arise in the analysis of forms such as *took* 'take.PST', *sang* 'sing.PST' and *went* 'go.PST'. "How shall we describe, now, the preterit form *took*? [...] Either *took* is one morpheme or it is two morphemes; the possibility of it being more than two may be neglected as improbable" (Bloch, 1947, p.400). The solution Bloch adopts is that *took* is two morphemes: but which morphemes, exactly? The reasoning is the following: "Some verbs [...] have a base with two different morpheme alternants: one that appears when the base is used alone and before certain of the inflectional suffixes, another that appears before certain other suffixes" (Bloch, 1947, p.404). *Take* is one of these verbs: in Bloch's analysis, the lexical morpheme *take* appears as

---

property 'past' is expressed both by the vowel -*o*- in the stem — vs. -*e*- in the present *sell* — and by the suffix -*d*).

162

the "alternant" *take* before the suffixes -*s* '3SG.PRS', -*en* 'PST.PTCP' and -*ing* 'gerund', while it appears as *took* before the suffix -Ø 'past'. This analysis, which makes crucial appeal to a zero morph, was soon criticized: Nida (1948, p.415) observed that such an analysis "means assigning a meaning-difference to certain covert elements rather than to overt distinctions" and continued: "it appears to me as strikingly contradictory to treat overt distinctions as meaningless and covert distinctions as meaningful" (Nida, 1948, p.415); even stronger criticism was expressed by Haas (1957, pp.34-35), who wrote: "It seems perverse to regard the very obvious phonemic difference between *went* and *go* as irrelevant to the semantic difference between the two expressions, and to suppose instead that the presence of some elusive zero suffix in *went*, as against the still more elusive absence of such from *go*, could serve to make the distinction". These criticisms are echoed more than half a century later by Anderson (2015, p.21), who writes, about *sang*: "We might say that the past tense morpheme here has a zero allomorph, and that *sang* is a predictable allomorph of *sing* that appears before this past tense zero. [...] it involves saying that the thing we cannot see, the zero, is what signals that the verb is past tense, while the thing we can see, the vowel change, is analyzed as a mechanical concomitant of this. The result does not correspond to any plausible intuition about how form and content are related". Anderson (2015, p.20) also observes that "describing this situation by appeal to a 'zero morph' does not provide a solution to the problem, but only a name for it". However, Bloch's analysis is still advocated by some contemporary "lexical" theories, such as the lexical-realizational Distributed Morphology (DM; Halle & Marantz, 1993): a DM analysis of *sang* maintains that the -Ø 'past' morpheme competes with -*ed* 'past'; since -Ø 'past' subcategorizes for a smaller class of verbs than -*ed* 'past' (which is the default suffix for the expression of the property 'past'), it is selected by virtue of Pāṇini's principle (also known as Elsewhere Condition); this zero suffix then triggers a rule of readjustment that changes the lexical morpheme's vowel (obtaining *sang* from *sing*). Stump (2001, p.10) observes that in DM analyses "again and again, both within and across languages, a default affix is overridden by an empty affix whose presence triggers a readjustment rule; this recurrent pattern is portrayed not as the consequence of any overarching principle, but as the accidental effect of innumerable piecemeal stipulations in the lexicon of one language after another. If one searched the languages of the world for a class of overt and phonologically identical affixes having the same sort of distribution that Halle and Marantz must logically attribute to their proposed class of empty affixes, one would inevitably come back empty-handed".[3] The fact that lexical and incremental theories force the

---

[3] Stump uses "empty affix" here rather than "zero affix": this is unfortunate, since "empty affix" is a term usually employed to designate the mirror-image of zero affixes, i.e. affixes that have overt phonological form but no meaning, such as thematic

analyst to posit dubious entities such as zero affixes is then considered as proof of the inadequacy of such theories.

A further argument against lexical and incremental theories is the fact that, in describing the meaning content of an inflected forms as contributed by the several morphemes that make it up, a choice is forced to distinguish between "properties of content and properties of context" (Stump, 2001, p.10). For example, in Italian past perfective forms such as *a m a i* 'love.PST.PFV.1SG', *te m e i* 'fear.PST.PFV.1SG', *sentii* 'hear.PST.PFV.1SG' a morpheme which has the signifier -*i* can be recognized; the meaning of this morpheme could be described either as '1SG' in the context 'PST.PFV' (since '1SG' is expressed by -*o* in other contexts, e.g. in the present indicative forms *a m o* 'love.PRS.IND.1SG', *t e m o* 'fear.PRS.IND.1SG', *s e n t o* 'hear.PRS.IND.1SG') or as 'PST.PFV.1SG'. The first option analyzes '1SG' as the content of -*i*, and 'PST.PFV' as the context in which -*i* '1SG' can be inserted, while the second option analyzes the bundle 'PST.PFV.1SG' as the content of -*i*. In Stump's (2001, p.11) words, "The problem is that there is no universally applicable criterion which determines whether a property belongs to an affix's content or to the context for which it subcategorizes". Lexical, morpheme-based theories force us to decide between the two analyses; in inferential-realizational theories, instead, there is no need for this decision. Such a theory would generate forms such as Italian *vidi* 'see.PST.PFV.1SG' or *amai* 'love.PST.PFV.1SG' in the following way: first, a rule of stem selection would select the appropriate stem of the lexeme (Stem 5 *vid*- in the case of 'see', since the 'PST.PFV.1SG' cell belongs to the paradigm partition that selects Stem 5, and the default stem *ama*- in the case of 'love', where Stem 5 is re-indexed to the default stem)[4]; then, a realization rule would insert -*i* in the context 'PST.PFV.1SG'; nothing in this series of operations specifies which portion of the signifier carries which portion of the meaning, but together these rules specify that *vidi* is 'see.PST.PFV.1SG' and *amai* is 'love.PST.PFV.1SG'.

Further arguments against lexical and incremental theories will not be reviewed for lack of space: the interested reader is referred to Stump (2001, 2016) for a very rich presentation and to Anderson (2015) for a short but very complete overview.

Inferential-realizational theories do not incur in the problems illustrated above: there is no need to posit zero morphs, since there is no requirement that a specific part of the signifier of an inflected form be the (only) overt exponent of a specific part of its meaning. A further advantage of inferential-realizational theories is that

---

vowels in Latin and Romance, interfixes, linking elements in compounds, and the like.
[4] The analysis of Italian verb stems and their distribution in paradigm partitions assumed here follows Pirrelli & Battista (2000), who applied to Italian verbs the concept of morphomic stems introduced by Aronoff (1994).

they easily accommodate non-concatenative means of exponence, such as tone or stress shifts, which can be introduced by realization rules but cannot be reduced to the concatenation of a specific morph.[5]

The exact working of an inferential-realizational theory (of which there are several varieties, including Stump's Paradigm Function Morphology and Anderson's A-morphous Morphology) is very technical, and cannot possibly be illustrated in this short review. But I hope to have shown that it deserves to be considered as a valid alternative to lexical-incremental, morpheme-based models. It is unfortunate that morpheme-based models are still considered as the only 'true' model of morphological analysis by most scholars who are not morphologists, while most prominent contemporary morphologists have come to the conclusion that "the term 'morpheme' has hindered rather than helped our understanding of how morphology works" (CarstairsMcCarthy, 2005, p.22), and regret the fact that "linguists continue with disconcerting regularity to regard analyses such as the decomposition of *unavoidable* into *un+avoid+able* as if it provided a perfectly general model of word structure" (Anderson, 2015, p.25).[6]

References

Anderson, S.R. (2015). The morpheme: Its nature and use. In M. Baerman (Ed.), *The Oxford Handbook of Inflection* (pp.11-33). Oxford: Oxford University Press.

Aronoff, M. (1994). *Morphology by itself: stems and inflectional classes.* Cambridge, MA: The MIT Press.

Bertinetto, P.M., Burani, C., Laudanna, A., Marconi, L., Ratti, D., Rolando, C. et al. (2 0 0 5). *Corpus e Lessico di Frequenza dell'Italiano Scritto (CoLFIS).*http://linguistica.sns.it/CoLFIS/Home.htm; https://www.istc.cnr.it/en/grouppage/colfis

Bloch, B. (1947). English verb inflection. *Language*, 23, 399-418.

Bonami, O., & Strnadová, J. (In press). Paradigm structure and predictability in derivational morphology. *Morphology*.

Burani, C. (2006). Morfologia: i processi. In A. Laudanna & M. Voghera (Eds.), *Il*

---

[5] Anderson (2015, p.32) clearly states that "Inferential-realization theories represent [...] the complete abandonment of the traditional concept of the morpheme".

[6] In the text I have dealt mainly with inflection, but many recent contributions argue that lexeme formation is as much amenable to paradigm-based, realizational theories as inflection; for two different but compatible approaches see Bonami & Strnadová (in press) and Masini & Audring (2019).

linguaggio. Strutture linguistiche e processi cognitivi (pp. 112-129). Roma-Bari: Laterza.

Carstairs-McCarthy, A. (2005). Basic terminology. In P. Štekauer & R. Lieber (Eds.), *Handbook of Word-formation* (pp. 5-23). Dordrecht: Springer.

Halle, M., & Marantz, A. (1993). Distributed morphology and the pieces of inflection. In K. Hale & S. J. Keyser (Eds.), *The view from building 20* (pp. 111-176). Cambridge, MA: The MIT Press.

Hockett C. F. (1954). Two Models of Grammatical Description, *Word*, 10, 210-231.

Masini, F. & Audring, J. (2019). Construction Morphology. In J. Audring & F. Masini (Eds.), *The Oxford Handbook of Morphological Theory* (pp. 365-389).Oxford: Oxford University Press.

Matthews, P.H. (1975). Sviluppi recenti nella morfologia. In J. Lyons–(Ed.), *Nuovi orizzonti della linguistica* (pp. 109-131). Torino: Einaudi.

Matthews, P.H. (1991). *Morphology. Second edition*. Cambridge, UK: Cambridge University Press.

Nida, E. (1948). The identification of morphemes. *Language*, 24, 414-441.

Pirrelli, V. & Battista, M. (2000). The paradigmatic dimension of stem allomorphy in Italian verb inflection. *Rivista di linguistica*, 12, 307-380.

Stump, G.T. (2001). *Inflectional morphology. A theory of paradigm structure*. Cambridge, UK: Cambridge University Press.

Stump, G.T. (2016). *Inflectional paradigms: Content and form at the syntax-morphology interface*. Cambridge, UK: Cambridge University Press.

Thornton, A. M. (1999). Diagrammaticità, uniformità di codifica e morfomicità nella flessione verbale italiana. In in P. Benincà, A.M. Mioni, & L. Vanelli (eds.), *Fonologia e morfologia dell'italiano e dei dialetti d'Italia* (pp. 483-502). Roma: Bulzoni.